National Conference on Contemporary Issues in High Energy Physics and Cosmology

(NC-HEPC 2013)

Journal of Physics: Conference Series Volume 481

Gauhati, India
12-14 February 2013

Editor:

Kalpana Bora

ISBN: 978-1-63266-703-8
ISSN: 1742-6588

Printed from e-media with permission by:

Curran Associates, Inc.
57 Morehouse Lane
Red Hook, NY 12571

Some format issues inherent in the e-media version may also appear in this print version.

Printed by Curran Associates, Inc. (2014)

For permission requests, please contact the Institute of Physics
at the address below.

Institute of Physics
Dirac House, Temple Back
Bristol BS1 6BE UK

Phone: 44 1 17 929 7481
Fax: 44 1 17 920 0979

techtracking@iop.org

Additional copies of this publication are available from:

Curran Associates, Inc.
57 Morehouse Lane
Red Hook, NY 12571 USA
Phone: 845-758-0400
Fax: 845-758-2634
Email: curran@proceedings.com
Web: www.proceedings.com

TABLE OF CONTENTS

Partons at Small *x* ... 1
D.K. Choudhury

New Frontiers in Neutrino Physics ... 6
N. Nimai Singh

Is the Scalar Higgs Boson Observed at the LHC, A Supersymmetric Higgs Boson or a Standard Model One? ... 13
Subrata Chakraborty

Dark Energy Model Building and Observational Issues ... 18
Anjan A. Sen

Bounds on Sterile Neutrino Parameters from Reactor Experiments ... 22
Kalpana Bora

Universal Extra Dimensions: Life with BLKTs ... 28
Anindya Datta, Ujjal Kumar Dey, Amitava Raychaudhuri, Avirup Shaw

A Discrepancy Observed in the Dipole Anisotropy in the Radio Sky at 1.4 GHz and that in the CMBR - A Threat to the Cosmological Principle? ... 34
Ashok K. Singal

Deflection of Light Due to Rotating Mass - A Comparison Among the Results of Different Approaches ... 39
Sarani Chakraborty, Asoke kumar Sen

***f(R)* Gravity Cosmology in Scalar Degree of Freedom** ... 43
Umananda Dev Goswami, Kabita Deka

Gravitational Redshift in Kerr Field ... 47
Anuj Kumar Dubey, Asoke Kumar Sen

Impact of Interacting Dark Sector on Cosmic Microwave Background Radiation ... 51
M.M. Verma

Non-Gaussian Signatures Arising from Warm in Inflation Driven by Geometric Tachyon ... 55
Anindita Bhattacharjee, Atri Deshamukhya

Study of IGM Through High Energy Radiation from Blazar ... 59
Jayashri Medhi, H.L. Duorah, A.G. Barua, K. Duorah

Thermodynamics of BTZ Black Hole and Entanglement Entropy ... 63
Dharm Veer Singh, Sanjay Siwach

Thermodynamic Origin of Interaction in the Cosmological Tachyonic Field ... 67
S.D. Pathak, M.M. Verma

Neutrino Masses and Mixings Using Updated Values of Running Quark and Lepton Masses ... 71
Kalpana Bora, Gayatri Ghosh

Discovery Reach of CP Violation in Neutrino Oscillation Experiments with Standard and Non-Standard Interactions ... 75
Zini Rahman, Arnab Dasgupta, Rathin Adhikari

Probing the Type-III Seesaw Model Through $e^+e^- \longrightarrow \Sigma^+\Sigma^-$ at ILC ... 79
Sumit K. Garg, Deepanjali Goswami, P. Poulose

Resolving Octant Degeneracy at LBL Experiment by Combining Daya Bay Reactor Setup ... 83
Kalpana Bora, Debajyoti Dutta

An Analysis of Isgur-Wise Function of Heavy-Light Mesons within a Higher Dimensional Potential Model Approach ... 87
Sabyasachi Roy, D.K. Choudhury

An Improved Analysis of Neutrino-Nucleon Interaction Cross-Section at Ultra High Energy ... 91
Pijush Kanti Dhar, D.K. Choudhury

Comments on the Perturbation of Cornell Potential in a QCD Potential Model ... 95
D.K. Choudhury, Krishna Kingkar Pathak

Non-Singlet Structure Function F_2^{NS} (x,t) in DGLAP Approach ... 99
Neelakshi N.K. Borah, D.K. Choudhury, P.K. Sahariah

Sea-Quark and Gluonic Contributions to Strange Baryons and Its Properties ... 103
M. Batra, A. Upadhyay

Self-Similarity Based Model of Double Parton Distribution Functions at LHC 107
Akbari Jahan, D.K. Choudhury
Solution of Nonlinear Gribov-Levin-Ryskin-Mueller-Qiu Evolution Equation for Gluon Distribution Function 111
Mayuri Devee, J.K. Sarma
Exactly Solvable Central Potentials Related to Romanovski Polynomials 115
Nabaratna Bhagawati
Warm Dust-Acoustic Shocks 119
Rajita Goswami, Murchana Khusroo, Madhurjya P. Bora
Author Index

Preface

The idea of holding a "National Conference in High Energy Physics and Cosmology" at the Physics Department of Gauhati University, was conceived in 2011, during a discussion with Professor Nimai Singh. In 2012 we started the process, and in June 2012, I wrote to some of the internationally reputed researchers in India, for delivering the invited talks. After getting consent from a good number of them, we launched our website at http://www.gauhati.ac.in/nchepc and, over time, everything seemed to fall into place!

The main objective of the conference was to provide a platform to young researchers in High Energy Physics and Cosmology from the north-eastern part of India, to interact with internationally renowned researchers and experts in the field from all over the country, and to present their research works to a wider audience. There were about 80 participants and about 60 papers were presented in various technical sessions. I would like to express my sense of gratitude to all the invited speakers who attended the conference, gave invited talks and interacted with young researchers. I would also like to thank all the participants for their contribution in various technical sessions. Sincere thanks also go to various funding agencies.

All the members of the Organizing Committee, who have contributed in various capacities, for the successful organization of the conference, deserve sincere thanks. I thank the patron of the Organizing Committee, Professor O K Medhi, former Vice Chancellor and Dr Mridul Hazarika, Vice Chnacellor of Gauhati University. People who deserve a special mention are Professor N Nimai Singh, the president of the Organising Committee, Professor Madhurjya P Bora for taking into account the conference website, abstract and paper submission, and publication of this proceedings, Professor Kalyanee Boruah and Professor A Gohain Barua, Dr Banty Tiru, Dr Manas Kalita, and Mr Sanjeev Kalita for their co-operation. And finally, I would like to thank our research scholars Nabaratna, Debajyoti, Shubhankar, Akbari, Chandan, Tapasi, Gayatri, Tapan, Kuldeep, Mintu, Runmi, Amir, Baishali, Nilakshi, Mousumi, and many others, who have really worked very hard towards the smooth running of the conference.

It indeed has been a memorable experience to hold this conference, though the task seemed to be challenging in the beginning. I hope the fruitful interactions among invited speakers and young researchers will turn out be beneficial for the future career of the latter, and to build prospective collaborations, in the field.

Dr Kalpana Bora, *Editor-in-Chief*
Gauhati University
28-11-2013

Journal of Physics: Conference Series **481** (2014) 011001 doi:10.1088/1742-6596/481/1/011001

Message from the Vice Chancellor, Gauhati University

It gives me immense pleasure to know that the proceedings of the NC-HEPC 2013, held during the month of February, 2013 is being published in the Journal of Physics, Conference Series, by the IOP (Institute of Physics, UK). Deliberations in the conference, I am sure, have added new ideas to the pool of the exiting knowledge in the field.

This is a highly commendable effort by the Physics Department of Gauhati University for which they deserve due appreciation from all concerned. I hope that the journal to be published will be immensely beneficial particularly to the students and teachers of physics.

I wish the Organising Committee will endeavour to organise such events in near future.

Dr Mridul Hazarika
Vice-Chancellor, Gauhati University
28-11-2013

Journal of Physics: Conference Series **481** (2014) 011001 doi:10.1088/1742-6596/481/1/011001

Committees

Organising Committee

Professor O K Medhi, former Vice Chancellor, Gauhati University (Patron)
Professor N Nimai Singh (Chairperson)
Dr. Kalpana Bora (Convener)
Professor M P Bora (Secretary)
Mr. Sanjeev Kalita (Assistant Secretary)

Invited Speakers

Professor D P Roy, TIFR, Mumbai
Professor Naba K Mondal, TIFR, Mumbai
Professor D K Choudhury, Gauhati University
Professor Utpal Sarkar, PRL, Ahmedabad
Professor Raj Gandh, HRI, Allahabad
Professor Amol Dighe, TIFR, Mumbai
Professor Nita Sinha, IIMSc, Chennai
Professor S Chakraborty, Vishwa Bharati University, Santiniketan
Dr A A Sen, JMI, Delhi
Dr Prakash Mathews, SINP, Kolkata
Dr Kalpana Bora, Gauhati University
Dr A Dutta, Calcutta University

Sponsors

Gauhati University

DAE, Govt of India

भारत सरकार
विज्ञान और प्रौद्योगिकी मंत्रालय
विज्ञान और प्रौद्योगिकी विभाग

GOVERNMENT OF INDIA
Ministry of Science and Technology
Department of Science and Technology
Technology Bhavan, New Mehrauli Road
New Delhi - 110016

DST, Govt of India

ASTEC, Govt of Assam

TIFR

CSIR

Peer review statement

All papers published in this volume of *Journal of Physics: Conference Series* have been peer reviewed through processes administered by the proceedings Editors. Reviews were conducted by expert referees to the professional and scientific standards expected of a proceedings journal published by IOP Publishing.

Partons at small x

D K Choudhury
Department of Physics, Gauhati University, Guwahati-781014, India.
Physics Academy of the North-East, Guwahati-781014, India.
E-mail: dkc_phys@yahoo.co.in

Abstract. We give a brief summary of the work on partons at small Bjorken x pursued by us at Gauhati University in recent years. We use both conventional DGLAP method and not so conventional self-similarity inspired method. In DGLAP approach, we solve the QCD evolution equations approximately at small x by two methods- Lagrange's and method of characteristics. We apply them to both spin dependent and spin independent structure functions. In the second approach, we use the notion of self-similarity of fractal geometry to construct Transverse Momentum Dependent Parton Distributions (TMD's), integrated Parton Densities (PDF's) and determine them from HERA collider data. We then use them to compute fractions of momentum carried by quarks and gluons inside the proton and compare with QCD expectations. In view of topical importance of multi parton interactions at LHC at TeV scale, we also outline the construction of Double Parton Distributions (dPDF's) with self-similarity at small x_1, x_2.

1. Introduction

In deep inelastic e-p scattering, x denotes the fraction of momentum carried by partons in a nucleon. Small x ($x \ll 1$) partons were coined "wee" partons by Feynman more than forty years ago in pre QCD era, in late 1960's and early 1970's. Since then the subject attracted immense interest both theoretically and experimentally. Both in the experiments at lepton hadron collider "HERA" at DESY, Hamburg and the present LHC at CERN, Geneva, the subject has been attracting keen attention. [1-3]. As noted in the abstract, the paper summaries the specific aspects of small x partons. In section 2, we summarise the DGLAP based models while the section 3 is devoted to self-similarity based model of parton at small x.

2. DGLAP based models

In this approach, the Q^2 evolution of the parton distribution functions (PDF) or the structure functions are determined by a set of integro-differential equation in variable $t = \log \frac{Q^2}{\Lambda^2}$ known as DGLAP equation [4-6] named after the authors. In more recent years, several alternative evolution equations have also been suggested like BFKL [7], GLR [8], BK [9], and JIMWLK [10]. Each of them has varied degree of theoretical refinement compared to DGLAP. But at phenomenological level, DGLAP appears to be the most popular one both at HERA and LHC regimes.

As early as 1987 [11], a programme of obtaining approximate analytical solutions of these equations was initiated, focussing on small x in later years [12].

Neglecting the quark part and assuming factorization of x and Q^2, as

$$G(x, Q^2) = g(x).h(Q^2) \tag{1}$$

Sometimes back we obtained [13] the following gluon distributions:

$$G(x,t) = G(x,t_0)\left(\frac{t}{t_0}\right)^{p(x)} \tag{2}$$

with

$$p(x) = \frac{36}{25}\left\{\left(\frac{11}{12} - \frac{n_f}{18}\right) + ln(1-x) + I_g(x)\right\} \tag{3}$$

where the integral $I_g(x)$ is explicit function of x.

Using the standard approximate relation between the gluon and the longitudinal structure function $F_L(x,Q^2)$ [14], we also computed $F_L(x,Q^2)$ and results came close to data.

A new relation between the gluon and the slope of the structure function [15] derived by us

$$\frac{\partial F_2(x,Q^2)}{\partial \log Q^2} = \frac{5\alpha_s}{3\pi} G\left(\frac{4}{3}x\right) \tag{4}$$

was an improvement over the similar relation by Prytz [16] earlier.

DGLAP equations are differential equations in variable $t = \log\frac{Q^2}{\Lambda^2}$. Using a Taylor series approximation at small x, it was later shown [17] that these equations can be re-expressed as differential equations in two variables x and t of the form :

$$Q(x,t)\frac{\partial F_2(x,t)}{\partial t} + P(x,t)\frac{\partial F_2(x,t)}{\partial x} = R(x,t) \tag{5}$$

where $Q(x,t), P(x,t)$ and $R(x,t)$ are calculable.

There are two standard methods of solutions of such equations: (a) Lagrange's method [18] and (b) method of characteristics. [19] Our analysis with the Lagrange's method indicates that the solutions are not unique. While earlier study suggests [17,20],

$$F_2^{NS}(x,t) = F_2^{NS}(x,t_0)\left(\frac{t}{t_0}\right), \tag{6}$$

a general analysis in later years [21] indicates that equation(6) is a particular case of the most general solution at small x.

$$F_2^{NS}(x,t) = F_2^{NS}(x,t_0)\left(\frac{t}{t_0}\right)^{n(x,t)}, \qquad n > 0 \tag{7}$$

Our results achieved by the method of characteristics, on the other hand, suggests that the gluon density at small x [22] is given by

$$G(x,t) = G(x)\left(\frac{t}{t_0}\right)^{l(x,t)} \tag{8}$$

where, $l(x,t)$ is explicitly calculable. More recently, tensor structure function $b_1^d(x,Q^2)$ of the deuteron has been computed [23] using the method of characteristics and considering both NLO and NNLO effects using the relation

$$b_1^d(x,Q^2) = -\frac{3}{2}A_{zz}^d F_1^d(x,Q^2) \tag{9}$$

where the asymmetry factor A_{zz}^d is taken from HERMES data [24]. The results in ref [23] indicates that the predicted behaviour of b_1^d is in good agreement with experiment within the limits of uncertainty. A comparative analysis of the polarised structure function $g_1^{NS}(x,Q^2)$ and the unpolarised one $F_2^{NS}(x,Q^2)$ with the two methods have also been recently reported [25,26] besides a similar one by a new method: the method of successive approximation [27].

Table 1. Values of Q^2 for different values of allowed lower and upper bounds of quark and gluon momentum fractions [38]

Lower bounds of $\langle\hat{x}\rangle_q$	Upper bounds of $\langle\hat{x}\rangle_g$	Q^2 in GeV^2 $a = 1$	$a = 3.1418$
$\frac{9}{25}$	$\frac{16}{25}$	5.97×10^7	4.48×10^1
$\frac{3}{7}$	$\frac{4}{7}$	3.077×10^9	1.57×10^2
$\frac{15}{31}$	$\frac{16}{31}$	7.37×10^{10}	4.32×10^2
$\frac{1}{2}$	$\frac{1}{2}$	1.86×10^{11}	5.80×10^2
$\frac{9}{17}$	$\frac{8}{17}$	1.01×10^{12}	9.94×10^2
$\frac{9}{13}$	$\frac{4}{13}$	1.17×10^{16}	1.95×10^4
$\frac{3}{4}$	$\frac{1}{4}$	3.22×10^{17}	5.62×10^4
$\frac{15}{19}$	$\frac{4}{19}$	3.11×10^{18}	1.16×10^5
$\frac{9}{11}$	$\frac{1}{11}$	1.62×10^{19}	1.95×10^5
1	0	5.58×10^{23}	5.43×10^6

3. Self-similarity based models

Self-similarity is an inherent property of fractals [28]. Since 1980's, the notion of fractals has found its application in high energy physics through the self-similar nature of hadron multi-particle distribution [29,30]. Relevance of these ideas in deep-inelastic scattering was first noted by Dremin and Levtchenko [31]. Later, Lastovicka [32] suggested how fractality can be used to construct PDF's and structure function at small x. Choosing the relevant scaling variables as $\frac{1}{x}$ and $(1 + Q^2/Q_0^2)$, following form of PDF was obtained :

$$q_i(x, Q^2) = \frac{e^{D_0^i}\, Q_0^2\, x^{-D_2}}{1 + D_3 + D_1 ln\frac{1}{x}} \left[\left(\frac{1}{x}\right)^{D_1 ln\left(1+\frac{Q^2}{Q_0^2}\right)} \left(1 + \frac{Q^2}{Q_0^2}\right)^{D_3+1} - 1 \right] \tag{10}$$

where the parameters D_0^i, Q_0^2, D_1, D_2, D_3 are model parameters fitted from HERA data [33].

Later, an alternative parameters of the PDF was suggested [34] and applied to Ultra High Energy Neutrino-Nucleon collision [35].

The model of equation(10) can be used to calculate the fraction of momentum carried by quarks and gluons using the momentum sum rule [36] and assuming its validity for $0 < x < 1$. A recent analysis [37] indicates that in such a model, fraction of momentum of quarks $\langle x\rangle_q$ increases with Q^2 while that of gluons $\langle x\rangle_g$ decreases. At $Q^2 = 45$ GeV^2, on the other hand $\langle x\rangle_q \sim 1$ suggests the saturation of the momentum sum rule. In a more recent work [38] $\langle x\rangle_q$ and $\langle x\rangle_g$ were re-estimated for partons within the limited x-range $6.2 \times 10^{-7} \leq x \leq 10^{-2}$ [32]. It leads to lower bounds $\langle x\rangle_q$ and upper bounds $\langle x\rangle_g$ in the model [Table1]. In the table1, $a = 1$ and $a = 3.1418$ corresponds to integrally and fractionally charged partons. In this case, the momentum sum rule saturates at an energy scale beyond the reach of even ultra high energy neutrinos.

Let us now discuss the implication of the models at LHC. At hadron collider like LHC, each hadron is described as a collection of essentially free elementary constituents. The interactions between constituents belonging to different colliding hadrons are the seeds of multi-partonic interaction (MPI). Due to the composite nature of each hadron, it is possible to have multiple parton hard scattering: i.e. events in which two or more distinct hard parton scattering occur simultaneously in a single hadron-hadron collision. In such a case, double parton scattering (DPS) plays an important role. Experimentally, such DPS leads to four jet events at hadron colliders. CMS collaboration and ATLAS collaboration are already studying such processes [39].

In QCD, to study such DPS, one needs double PDFs (dPDFs) of each hadron and their evolution. Unlike conventional PDFs, DGLAP equations are generalised to double DGLAP [40] equations to study evolution of dPDFs.

Recently, we have suggested a model of dPDFs based on self-similarity at small x_1 and x_2 [41]. The model contains total thirteen parameters to be fitted from the data at LHC. It is also shown that the constructed dPDF does not factorise into the single PDFs in conformation with QCD expectation. It also satisfies the condition that at the kinematic boundary $x_1 + x_2 = 1$ (where x_1 and x_2 are the longitudinal fractional momentum of two interacting partons), the dPDF vanishes and its simplest form under plausible assumptions is

$$D_{ij}(x_1, x_2) \sim \left(\frac{1}{x_1}\right)^{D_1 \log \frac{1}{x_2}} \left(\frac{1-(x_1+x_2)}{x_1+x_2}\right)^{D_2} \tag{11}$$

where D_1 and D_2 are the model parameters to be fitted. Further work is under progress [42] to make the self-similarity based models compatible with DGLAP evolution as well as Froissart saturation.

4. Acknowledgments

I thank the organisers of the NC-HEPC, Gauhati University for inviting me to deliver the talk. I also thank the collaborators who contributed at several stages of the developments of the present work spending more than two decades. Finally I acknowledge the help of Ms Baishali Saikia in preparing the manuscript. This talk is dedicated to the memory of Prof S. A. S. Ahmed(1948-2012).

References

[1] L. Frankfurt, arXiv:hep-ph/0507286.
[2] D. A. Enteria, arXiv:hep-ex/0610062.
[3] N. P. Zotov, arXiv:hep-ph/1201.4144.
[4] V. N. Gribov and L. Lipatov, *Sov. J. Nucl. Phys.* **15**, 438 (1972).
[5] G. Altarelli and G. Parisi, *Nucl. Phys.* **B126**, 298 (1977).
[6] Y. L. Dokshitzer, *Sov. Phys. JETP* **46**, 641 (1977).
[7] V. S. Fadin, E. A. Kuraev and L. N. Lipatov, *Phys. Lett.* **B60** 50 (1975).
E. A. Kuraev, L. N. Lipatov and V. S. Fadin, *Sov. Phys. JETP* **44** 443 (1976).
I. I. Balitsky and L. N. Lipatov, *Sov. J. Nucl. Phys.* **28** 822 (1978).
[8] L. V. Gribov, E. M. Levin and M. G. Ryskin, *Phys. Rept.* 100 (1983).
[9] I. I. Balitsky, *Nucl. Phys.* **B463** 99 (1996).
Y. V. Kovchegov, *Phys. Rev.* **D60** 034008 (1999).
Y. V. Kovchegov, *Phys. Rev.* **D61** 074018 (2000).
[10] J. Jalilian-Marian, A. Kovner, A. Leonidov and H. Wignert, *Nucl. Phys.* **B504** 415 (1997).
E. Iancu, A. Leonidov and L. D. McLerran, *Phys. Lett.* **B510** 133 (2001).
[11] D. K. Choudhury and A. Saikia, *Pramana. J. Phys.* **29** 345 (1987).
[12] D. K. Choudhury and J. K. Sarma, *Pramana. J. Phys.* **38** 481 (1992).
[13] D. K. Choudhury and R. Deka, *Z. Phys.* **C75** 679 (1997).
R. Deka and D. K. Choudhury and A. Saikia, *Euro. Phys. J.* **C2** 302 (1998).
[14] A.M. Cooper Sarkar, G. Ingelman, K.R. Long, R.G. Roberts and D.H. Saxon *Z. Phys.* **C39** 281 (1988).
[15] K. Bora and D. K. Choudhury, *Phys. Lett.* **B354** 151 (1995).
[16] K. Prytz, *Phys. Lett.* **B311** 286 (1993).
[17] J. K. Sarma, D. K. Choudhury and G. K. Medhi, *Phys. Lett.* **B403** 139 (1997).
[18] I. N. Sneddon, *Elements of Partial differential Equation* (McGraw Hill, New York, NY USA, 1957).
[19] W. E. Williams, *Partial Differential Equations* (Clarendon Press, Oxford, UK 1980).
[20] D. K. Choudhury and Atri Deshamukhya, *Ind. J. Phys.* **75A** 573 (2001).
[21] D. K. Choudhury and P. K. Dhar, *Ind. J. Phys.* **81** 259 (2007).

[22] D. K. Choudhury, *P K Saharia, Pramana J. Phys.* **58** 599 (2002).
[23] Saiful Islam and D. K. Choudhury, *Euro. Phys. J.* **C72** 2257 (2012).
[24] A. Airapetian *et al., Phys. Rev. Lett.* **95** 242001 (2005).
[25] Neelakshi N. K. Bora, D. K. Choudhury and P. K. Saharia, *Pramana J. Phys.* **79** 833 (2012).
[26] Neelakshi N. K. Bora, D. K. Choudhury and P. K. Saharia, *Adv. High Energy Phys.* 2013 829803 (2013).
[27] D. K. Choudhury and Ranjit Choudhury, *Ind. J. Phys.* **86** 377 (2012).
[28] B. B. Mandelbroat, *Fractal Geometry of Nature*, W H Freman (New York 1982)
[29] I. M. Dremin, *Mod. Phys. Lett.* **A3** 1333 (1988).
[30] W. Florkowski and R. C. Hwa, *Phys. Rev.* **D43** 1548 (1991).
[31] I. M. Dremin and B. B. Levtchenko, *Phys. Lett.* **B262** 155 (1992).
[32] T. Lastovicka, *Euro. Phys. J.* **C24** 529 (2002).
[33] H1: C. Adloff *et al., Eur. Phys. J.* **C21**, 33-61 (2001), arXiv:hep-ex/0012053.
H1: S. Aid *et al., Z. Phys.* **C69**, 27 (1995), arXiv:hep-ex/9509001.
[34] D. K. Choudhury and Rupjyoti Gogoi, *Ind. J. Phys.* **80** 827 (2006).
[35] Rupjyoti Gogoi and D. K. Choudhury, *Int. J. Applied. Phys.* **2** 227 (2012).
[36] F. E. Close, *An Introduction to Quarks and Gluons* (Academic Press 1979), p.233.
[37] Akbari Jahan and D. K. Choudhury, *Mod. Phys. Lett.* **A27** 1250193 (2012), arXiv:1304.6882 [hep-ph].
[38] Akbari Jahan and D. K. Choudhury, *Mod. Phys. Lett.* **A28** 1350086 (2013), arXiv:1306.1891 [hep-ph].
[39] P. Bartalini *et al.*, arXiv:1111.0469 [hep-ph].
[40] A. M. Snigirev, *Phys. Rev.* **D68** 114012 (2003).
V. L. Korotkikh and A. M. Snigirev, *Phys. Lett.* **B594** 171 (2004).
[41] D. K. Choudhury and Akbari Jahan *IJMP* **28** 1350079 (2013), arXiv:1305.6180 [hep-ph].
[42] Baishali Saikia, D. K. Choudhury, Rupjyoti Gogoi and Akbari Jahan (Under Preparation).

New frontiers in neutrino physics

N Nimai Singh

Department of Physics, Gauhati University, Guwahati 781014, India

E-mail: nimai03@yahoo.com

Abstract. The origin of neutrino masses and mixings is an important topic in high energy physics. In the present work, two mixing schemes namely Bi-maximal (BL) and Tri-bimaximal (TBM) mixings which are derived from $\mu-\tau$ symmetric neutrino mass matrix, are examined after taking corrections from charged lepton sector. While tri-bimaximal mixing with appropriate charged lepton correction, is in good agreement with latest observations, the Bi-maximal mixing after taking charged lepton correction, fails to agree completely with observation. A new scheme called strict Bi-large (BL) mixing scheme is introduced in place of BM mixing scheme.The BL scheme works very well in complete agreement with data. Some of the unsettled issues in neutrino physics are outlined for further investigations.

1. Introduction

Recent observations [1] on neutrino oscillation indicate large values of $\sin^2\theta_{13}$ i.e, $\theta_{13} \sim \theta_c/\sqrt{2}$. The question being asked is: Is it accidental or manifestation of some underlying quark-lepton complementarity (QLC) relations? Further, the recent data [2] also shows certain indication of a mild deviation of θ_{23} mixing from the maximal $\tan^2\theta_{23} = 1.0$ i.e., $d_{23} = 0.5 - \sin^2\theta_{23} \sim 0.1$. Significant deviation of θ_{12} from tri-bimaximal mixing $\left(\tan^2\theta_{12} = 0.5\right)$ is an important issue in search for the nature of U_{PMNS} matrix. We have also witnessed the first gilimpse on the CP violating Dirac phase from the global fit, and a possibility of theoretical predictions for δ_{CP} is an exciting field of study. Since the leptonic mixing angles are relatively large, it is important to explore more detailed searches on large CP-violation effect in the leptonic sector.

2. CKM matrix in quark physics

We first describe the quark mixing matrix known as CKM matrix [3], for possible comparison with the corresponding leptonic mixing matrix known as PMNS matrix [4]. This connection is illustrated by a simple process like β decay. The β-decay process is represented as $n \rightarrow p+e^-+\overline{\nu}_e$. In term of quarks, it is replaced by $d \rightarrow u + e^- + \overline{\nu}_e$.

The correspondence between the mass-eigenstates and weak-flavour eigenstates, leads to the quark mixing matrix known as CKM matrix, and this is illustrated as follows,

$$\left.\begin{aligned} \begin{pmatrix} u & c & t \\ d & s & b \end{pmatrix} &\Rightarrow \begin{pmatrix} u & c & t \\ d' & s' & b' \end{pmatrix} \\ \begin{pmatrix} d' \\ s' \\ b' \end{pmatrix} &= U_{\text{CKM}} \begin{pmatrix} d \\ s \\ b \end{pmatrix} \end{aligned}\right\} \qquad (1)$$

where the CKM matrix is given by a 3×3 unitary matrix,

$$\mathbf{U}_{\text{CKM}} = \begin{pmatrix} U_{ud} & U_{us} & U_{ub} \\ U_{cd} & U_{cs} & U_{cb} \\ U_{td} & U_{ts} & U_{tb} \end{pmatrix}. \tag{2}$$

Parametrization [3] of CKM matrix can be done by the product of three rotational matrices: $\theta_{23}, \theta_{13}, \theta_{12}$ leading to $U_{\text{CKM}} = R_{23}R_{13}R_{12}$.

The final form of U_{CKM} is given by

$$\mathbf{U}_{\text{CKM}} = \begin{pmatrix} c_{12}c_{13} & s_{12}c_{13} & s_{13}e^{-i\delta} \\ -s_{12}c_{23} - c_{12}s_{23}s_{13}e^{i\delta} & c_{12}c_{23} - s_{12}s_{23}s_{13}e^{i\delta} & s_{23}c_{13} \\ s_{12}s_{23} - c_{12}c_{23}s_{13}e^{i\delta} & -c_{12}s_{23} - s_{12}c_{23}s_{13}e^{i\delta} & c_{23}c_{13} \end{pmatrix}. \tag{3}$$

In term of Wolfestein's parametrization [3], CKM matrix takes the form,

$$\mathbf{U}_{\text{CKM}} \approx \begin{pmatrix} 1 - \frac{\lambda^2}{2} & \lambda & A\lambda^3(\rho - i\eta) \\ -\lambda & 1 - \frac{\lambda^2}{2} & A\lambda^2 \\ A\lambda^3(1 - \rho - i\eta) & -A\lambda^2 & 1 \end{pmatrix} \tag{4}$$

where, $\lambda \approx 0.22, |A| \approx 0.9 \pm 0.18, \eta \ll 0, \tan\theta_{12} = |\frac{V_{us}}{V_{ud}}|, \tan\theta_{23} = |\frac{V_{cb}}{V_{tb}}|, \sin\theta_{13} = |V_{ub}|$; $s_{13} \approx 0, \sin\theta_{12} \approx \lambda \approx 0.22$; $\theta_{12} = \theta^c = 12.6° \pm 0.1°$.

3. Neutrino mixing from neutrino oscillation

In order to explain the neutrino mixings derived from neutrino oscillation phenomenon, we hypothesise the existence of mass eigenstate as well as flavour eigenstate in leptonic sector in analogy with quark sector. This is illustrated as follows,

$$\begin{pmatrix} e^- & \mu^- & \tau^- \\ \nu_1 & \nu_2 & \nu_3 \end{pmatrix} \Rightarrow \begin{pmatrix} e^- & \mu^- & \tau^- \\ \nu_e & \nu_\mu & \nu_\tau \end{pmatrix}$$

$$\begin{pmatrix} \nu_e \\ \nu_\mu \\ \nu_\tau \end{pmatrix} = U_{\text{PMNS}} \begin{pmatrix} \nu_1 \\ \nu_2 \\ \nu_3 \end{pmatrix}$$

which leads to the PMNS mixing matrix,

$$\mathbf{U}_{\text{PMNS}} = \begin{pmatrix} U_{e1} & U_{e2} & U_{e3} \\ U_{\mu 1} & U_{\mu 2} & U_{\mu 3} \\ U_{\tau 1} & U_{\tau 2} & U_{\tau 3} \end{pmatrix} = R_{23}R_{13}R_{12}. \tag{5}$$

Though the parametrization is through three rotations like CKM matrix, the values of three angles and phase are different in both cases. In case of PMNS matrix, the three angles are relatively large. They are extracted from neutrino oscillation experiments. Probability of neutrino oscillation from $\nu_e \to \nu_\mu$ is given by

$$P(t)_{\nu_e \to \nu_\mu} = \sin^2(2\theta_{12}) \sin^2 \left[1.267\Delta m^2_{21} \frac{L}{E}\right]$$

where, θ_{12} is the solar mixing angle, L is in metre, E in MeV, Δm^2_{21} in eV^2. In the structure of U_{PMNS}, unlike CKM matrix, there are two complex Majorana phases represented by

$$P = \begin{pmatrix} e^{i\theta_1} & 0 & 0 \\ 0 & e^{i\theta_2} & 0 \\ 0 & 0 & 1 \end{pmatrix}. \tag{6}$$

if neutrino masses are Majorana type. Majorana fermions are self-conjugate particles and neutrino could satisfy this condition. Thus $U_{\rm PMNS} = R_{23}R_{13}R_{12} \cdot P$. The structure of $U_{\rm PMNS}$ takes the following form,

$$U_{\rm PMNS} = \begin{pmatrix} c_{12}c_{13} & s_{12}c_{13} & s_{13}e^{-i\delta} \\ -s_{12}c_{23} - c_{12}s_{23}s_{13}e^{i\delta} & c_{12}c_{23} - s_{12}s_{23}s_{13}e^{i\delta} & s_{23}c_{13} \\ s_{12}s_{23} - c_{12}c_{23}s_{13}e^{i\delta} & -c_{12}s_{23} - s_{12}c_{23}s_{13}e^{i\delta} & c_{23}c_{13} \end{pmatrix} \cdot P \qquad (7)$$

In order to construct the PMNS mixing matrix, we have to get the diagonalizing matrices from neutrino mass matrix as well as charged lepton mass matrix in analogy to the construction of CKM matrix. Two conventions [5-8] of diagonalization are adopted. We follow a convention which gives the following relations: $U_{\rm CKM} = U^{\dagger}_{uL}U_{dL}$, $m_e = U_{eL}m_e^{\rm diag}U^{\dagger}_{eR}$, $U_{\rm PMNS} = U_e^{\dagger}U_{\nu}$ and $m_{\nu} = U^{*}_{\nu}m_{\nu}^{\rm diag}U^{\dagger}_{\nu}$. The neutrino mass matrix is expressible in the basis where charged lepton mass matrix is diagonal, $m'_{\nu} = \left(U^{\dagger}_{eL}m_{\nu}U_{eL}\right)$.

4. Extracting $U_{\rm PMNS}$ matrix

The usual ways of getting neutrino mass matrix is from left-right symmetric GUT models such as SO(10) GUT, through celebrated seesaw formula [9], $m_{LL} = m_{LR}m^{-1}_{RR}m^{\dagger}_{LR}$. Another possible source of neutrino mass matrix is from certain discrete symmetries like $\mu-\tau$ symmetry [10]. The $\mu-\tau$ symmetric mass matrix leads to the following PMNS mixing matrix where solar mixing angle is arbitrary,

$$U_{\rm PMNS} = \begin{pmatrix} c_{12} & s_{12} & 0 \\ \frac{-s_{12}}{\sqrt{2}} & \frac{c_{12}}{\sqrt{2}} & \frac{1}{\sqrt{2}} \\ \frac{s_{12}}{\sqrt{2}} & \frac{-c_{12}}{\sqrt{2}} & \frac{1}{\sqrt{2}} \end{pmatrix}. \qquad (8)$$

When $s_{12} = \frac{1}{\sqrt{2}}$, $\tan^2\theta_{12} = 1.0$ leads to Bimaximal [11] mixing matrix (BM)

$$U_{BM} = \begin{pmatrix} \frac{1}{\sqrt{2}} & \frac{1}{\sqrt{2}} & 0 \\ -\frac{1}{2} & \frac{1}{2} & \frac{1}{\sqrt{2}} \\ \frac{1}{2} & -\frac{1}{2} & \frac{1}{\sqrt{2}} \end{pmatrix}. \qquad (9)$$

When $s_{12} = \frac{1}{\sqrt{3}}$, $\tan^2\theta_{12} = 0.5$ leads to tri-bimaximal [12] mixing matrix (TBM)

$$U_{\rm TBM} = \begin{pmatrix} \sqrt{\frac{2}{3}} & \frac{1}{\sqrt{3}} & 0 \\ -\sqrt{\frac{1}{6}} & \frac{1}{\sqrt{3}} & \frac{1}{\sqrt{2}} \\ \sqrt{\frac{1}{6}} & -\frac{1}{\sqrt{3}} & \frac{1}{\sqrt{2}} \end{pmatrix}. \qquad (10)$$

The validity of both BM as well as TBM will be examined with appropriate charged lepton corrections. This is the central issue of the present investigation.

5. Charged lepton correction to BM mixing with CP phase

To rescue the Bimaximal mixing (BM), we follow the procedure of taking charged lepton correction in two ways [13] for introducing the Dirac CP Phase. Thus

A. Dirac CP Phase through Charged lepton mixing matrix (U_{el}) :

$$U_{el} = \begin{pmatrix} \tilde{c}_{12} & \tilde{s}_{12}e^{-i\phi} & 0 \\ -\tilde{s}_{12}e^{i\phi} & \tilde{c}_{12} & 0 \\ 0 & 0 & 1 \end{pmatrix} \qquad (11)$$

and this leads to neutrino mixing matrix ($U_{\rm PMNS} = U_{eL}^{\dagger} U_{BM}$),

$$U_{\rm PMNS} = \begin{pmatrix} \frac{1}{\sqrt{2}}(\tilde{c}_{12} + \frac{\tilde{s}_{12}}{\sqrt{2}} e^{i\phi}) & \frac{1}{\sqrt{2}}(\tilde{c}_{12} - \frac{\tilde{s}_{12}}{\sqrt{2}} e^{i\phi}) & -\frac{\tilde{s}_{12}}{\sqrt{2}} e^{i\phi} \\ -\frac{1}{2}(\tilde{c}_{12} - \sqrt{2}\tilde{s}_{12} e^{-i\phi}) & \frac{1}{2}(\tilde{c}_{12} + \sqrt{2}\tilde{s}_{12} e^{-i\phi}) & \frac{\tilde{c}_{12}}{\sqrt{2}} \\ \frac{1}{2} & -\frac{1}{2} & \frac{1}{\sqrt{2}} \end{pmatrix}. \quad (12)$$

B. Dirac CP phase through neutrino mass sector:

$$U_{\rm PMNS} = U_{el}^{\dagger} R_{23} Diag\left(e^{i\delta}, 1, e^{-i\delta}\right) R_{12}. \quad (13)$$

Using $\theta_{23} = \frac{\pi}{4}$ and $\theta_{12} = \frac{\pi}{4}$ the complete PMNS mixing matrix is given by:

$$U_{\rm PMNS} = \begin{pmatrix} \frac{1}{\sqrt{2}}(\tilde{c}_{12} e^{i\delta} + \frac{\tilde{s}_{12}}{\sqrt{2}}) & \frac{1}{\sqrt{2}}(\tilde{c}_{12} e^{i\delta} - \frac{\tilde{s}_{12}}{\sqrt{2}}) & -\frac{\tilde{s}_{12}}{\sqrt{2}} e^{-i\delta} \\ -\frac{1}{2}(\tilde{c}_{12} - \sqrt{2}\tilde{s}_{12} e^{i\delta}) & \frac{1}{2}(\tilde{c}_{12} + \sqrt{2}\tilde{s}_{12} e^{i\delta}) & \frac{\tilde{c}_{12}}{\sqrt{2}} e^{-i\delta} \\ \frac{1}{2} & -\frac{1}{2} & \frac{1}{\sqrt{2}} e^{-i\delta} \end{pmatrix}. \quad (14)$$

$$Amethod: \quad (15)$$

$$J_{\rm CP}^{BM} = \frac{1}{4\sqrt{2}} \sin\tilde{\theta}_{12} \cos\tilde{\theta}_{12} \sin\phi \quad (16)$$

$$(17)$$

$$tan^2\theta_{12} = \frac{2 - \tilde{s}_{12}^2 - 2\sqrt{2}\tilde{c}_{12}\tilde{s}_{12}\cos\phi}{2 - \tilde{s}_{12}^2 + 2\sqrt{2}\tilde{c}_{12}\tilde{s}_{12}\cos\phi} \quad (18)$$

$$Bmethod: \quad (19)$$

$$J_{\rm CP}^{BM} = \frac{1}{4\sqrt{2}} \sin\tilde{\theta}_{12} \cos\tilde{\theta}_{12} \sin\delta \quad (20)$$

$$(21)$$

$$tan^2\theta_{12} = \frac{2 - \tilde{s}_{12}^2 - 2\sqrt{2}\tilde{c}_{12}\tilde{s}_{12}\cos\delta}{2 - \tilde{s}_{12}^2 + 2\sqrt{2}\tilde{c}_{12}\tilde{s}_{12}\cos\delta} \quad (22)$$

$$(23)$$

Both methods give the same expressions. For $\delta \neq 0$ the prediction on $\tan^2\theta_{12}$ cannot accomodate global data. However the prediction on atmospheric angle agrees with data [13].

Using $\sin\tilde{\theta}_{12} = \sqrt{2}\sin\theta_{13}$ along with the approximation $\cos\tilde{\theta}_{12} \approx 1$ gives

$$J_{\rm CP}^{BM} \approx \frac{1}{4} \sin\theta_{13} \sin\delta.$$

6. Charged lepton correction to Tri-bimiximal mixing with and without Dirac CP phase

In order to rescue the TBM we suggest the following charged leptonic mixing matrix satisfying unitarity condition

$$\tilde{U}_{el} = \begin{pmatrix} 1 - \frac{\lambda^2}{4} & -\frac{\lambda}{2} & -\frac{\lambda}{2} \\ \frac{\lambda}{2} & 1 - \frac{\lambda^2}{8} & -\frac{\lambda^2}{8} \\ \frac{\lambda}{2} & -\frac{\lambda^2}{8} & 1 - \frac{\lambda^2}{8} \end{pmatrix} \quad (24)$$

Using $U_{\rm PMNS} = \tilde{U}_{el}^{\dagger} U_{\rm TBM}$ we get the complete PMNS mixing matrix,

$$U_{\rm PMNS} = \begin{pmatrix} \sqrt{\frac{2}{3}}(1-\frac{\lambda^2}{4}) & \frac{1}{\sqrt{3}}(1-\frac{\lambda^2}{4}) & \frac{\lambda}{\sqrt{2}} \\ -\frac{1}{\sqrt{6}}(1+\lambda) & \frac{1}{\sqrt{3}}(1-\frac{\lambda}{2}) & \frac{1}{\sqrt{2}}(1-\frac{\lambda^2}{4}) \\ \frac{1}{\sqrt{6}}(1-\lambda) & -\frac{1}{\sqrt{3}}(1+\frac{\lambda}{2}) & \frac{1}{\sqrt{2}}(1-\frac{\lambda^2}{4}) \end{pmatrix}. \tag{25}$$

However in order to achieve $\tan^2\theta_{23} < 1$, we modify the charged lepton mixing matrix $\tilde{U}_{el}$ as

$$U_{el}^{\dagger} = \tilde{R}_{23}^{\dagger} \tilde{U}_{el}^{\dagger}$$

where $\tilde{R}_{23}$ is given by

$$\tilde{R}_{23} = \begin{pmatrix} 1 & 0 & 0 \\ 0 & \tilde{c}_{23} & \tilde{s}_{23} \\ 0 & -\tilde{s}_{23} & \tilde{c}_{23} \end{pmatrix}, S_{23}^2 = A\lambda^2 \approx 0.041. \tag{26}$$

This leads to the following predictions

$$\tan^2\theta_{12} = 0.5 \tag{27}$$

$$\tan^2\theta_{23} = \left(\frac{1-\tan\tilde{\theta}_{23}}{1+\tan\tilde{\theta}_{23}}\right)^2 < 1 \tag{28}$$

$$\sin\theta_{13} = \frac{\lambda}{\sqrt{2}} \approx 0.146. \tag{29}$$

In order to incorporate Dirac type CP effect in TBM scheme, we use [13,14]

$$U_{TBC} = \begin{pmatrix} \sqrt{\frac{2}{3}}(1-\frac{\lambda^2}{4}) & \frac{1}{\sqrt{3}}(1-\frac{\lambda^2}{4}) & \frac{\lambda}{\sqrt{2}}e^{-i\delta} \\ -\frac{1}{\sqrt{6}}(1+\lambda e^{i\delta}) & \frac{1}{\sqrt{3}}(1-\frac{\lambda}{2}e^{i\delta}) & \frac{1}{\sqrt{2}}(1-\frac{\lambda^2}{4}) \\ \frac{1}{\sqrt{6}}(1-\lambda e^{i\delta}) & -\frac{1}{\sqrt{3}}(1+\frac{\lambda}{2}e^{i\delta}) & \frac{1}{\sqrt{2}}(1-\frac{\lambda^2}{4}) \end{pmatrix} \tag{30}$$

where the complete PMNS mixing matrix is given by $U_{\rm PMNS} = \tilde{R}_{23}^{\dagger} U_{TBC}$ with phase. Here the prediction on $\tan^2\theta_{12}, \tan^2\theta_{23}, \sin^2\theta_{13}$ are free from CP phase δ. And we also obtain

$$J_{\rm CP}^{TB} \approx \frac{1}{3\sqrt{2}} \sin\theta_{13} \sin\delta.$$

The desired deviation of atmospheric angle from maximal value is obtained. However the deviation of solar angle from TBM value is not possible here to explain the data.

6.1. A new approach with the changing of the position of phase in TBM case

Following the suggestion from Fritzsch and Xing [15], U_{13} is replaced by U'_{13}, resulting a new TBM matrix $U'_{\rm TBM}$,

$$U_{13} = \begin{pmatrix} c_{13} & 0 & s_{13}e^{-i\delta} \\ 0 & 1 & 0 \\ -s_{13}e^{i\delta} & 0 & c_{13} \end{pmatrix} \tag{31}$$

$$U'_{13} = \begin{pmatrix} c_{13}e^{i\delta} & 0 & s_{13} \\ 0 & 1 & 0 \\ -s_{13} & 0 & c_{13}e^{-i\delta} \end{pmatrix} \tag{32}$$

$$U'_{\rm TBM} = \begin{pmatrix} \sqrt{\frac{2}{3}}e^{i\delta} & \frac{1}{\sqrt{3}}e^{i\delta} & 0 \\ -\frac{1}{\sqrt{6}} & \frac{1}{\sqrt{3}} & \frac{1}{\sqrt{2}}e^{-i\delta} \\ \frac{1}{\sqrt{6}} & -\frac{1}{\sqrt{3}} & \frac{1}{\sqrt{2}}e^{-i\delta} \end{pmatrix} \tag{33}$$

After taking CKM type charged lepton correction [16] we get

$$U_{\rm PMNS} = U_l^\dagger U'_{\rm TBM} = U^\dagger_{\rm CKM} U'_{\rm TBM}.$$

which predicts the best fit value $\tan^2\theta_{12} = 0.47$ and $J_{\rm CP} = 0.0350$ for $\cos\delta \approx 0.147$. The prediction on atmospheric angle is in good agreement [16] with observation. If $\cos\delta = 1$ we get only $\tan^2\theta_{12} \approx 0.232$ which is very less as compared to data. The predictions of other two angles are unaffected by the phases i.e, $\sin^2\theta_{13} = 0.024$ and $\tan^2\theta_{23} = 0.80$. The CP violating parameter $J_{\rm CP}$ is also predicted with right order [16].

7. Strict Bi-large mixing with charged lepton correction

The problem in Bi-maximal mixing to produce correct solar mixing, is rectified with the ansatz of strict Bi-Large mixing [17] having the following choices,

$$\sin\theta_{13} = \lambda_c, \quad \sin\theta_{12} = 3\lambda_c, \quad \sin\theta_{23} = 3\lambda_c \quad . \tag{34}$$

where $\lambda_c \approx 0.22$. Strict BL mixing matrix thus takes the form [17],

$$U_{BL} = \begin{pmatrix} \frac{3}{4}(1-\frac{\lambda_c^2}{2}) & \frac{\sqrt{7}}{4}(1-\frac{\lambda_c^2}{2}) & \lambda_c \\ -\frac{3\sqrt{7}}{16}(1+\lambda_c) & \frac{9}{16}(1-\frac{7}{9}\lambda_c) & \frac{\sqrt{7}}{4}(1-\frac{\lambda_c^2}{2}) \\ \frac{7}{16}(1-\frac{9}{7}\lambda_c) & -\frac{3\sqrt{7}}{16}(1+\lambda_c) & \frac{3}{4}(1-\frac{\lambda_c^2}{2}) \end{pmatrix} \tag{35}$$

After inclusion of CKM-type charged lepton correction,

$$U_{eL} = \begin{pmatrix} 1-\frac{\lambda_c^2}{2} & \lambda_c e^{i\delta_{cp}} & 0 \\ -\lambda_c e^{-i\delta_{cp}} & 1-\frac{\lambda_c^2}{2} & 0 \\ 0 & 0 & 1 \end{pmatrix} \tag{36}$$

we get the following predictions,

$$\sin^2\theta_{13} = 0.0245, \sin^2\theta_{12} = 0.3209, \sin^2\theta_{23} = 0.4533,$$

$$\delta_{\rm CP} = 0.2515\pi, J^{BL}_{\rm CP} \approx 0.0216$$

The predictions [17] of BL mixings on solar, atmospheric and Chooz mixing angles, are in agreement with recent data. Bilarge Mixing can be explained from certain underlying discrete symmetry in string theory.

8. Conclusions

To summarise, we have described the analogy between the CKM and PMNS matrix through the existence of both mass eigenstates and flavour eigenstates. The parametrizations in both cases are presented in terms of three rotations. In case of PMNS mixing matrix, the Bimaximal mixing matrix is nearly ruled out [13,17] even after charged lepton correction is included. Bilarge mixing matrix [17] or similar modification may be introduced to meet the challanges. Tri-bimaximal mixing is very much consistent with charged lepton correction, with or without phase contribution [13,16].

Some of the pending major issues on neutrino physics are - nature of neutrino mass: Dirac or Majorana nature [18]; pattern of absolute neutrino masses: normal, inverted or qusi-degenerate type [19]; possible existence of sterile neutrinos: a fourth state of neutrino flavour etc. The connection between the neutrinos and the Dark side [20] of the universe will continue to be an exciting research area [21].

Journal of Physics: Conference Series **481** (2014) 012002 doi:10.1088/1742-6596/481/1/012002

References

[1] DOUBLE-CHOOZ Collaboration, Abe Y et al. 2012 *Phys. Rev. Lett.* **108** 131801; DAYA-BAY Collaboration An F et al. 2012 *Phys. Rev. Lett.* **108** 171803; RENO Collaboration, Ahn J et al. 2012 *Phys. Rev. Lett.* **108** 191802

[2] Fogli G L et al. 2012 *Phys. Rev.* D **86** 013012; Boucenna S M et al. 2012 *Phys. Rev.* D **86** 051301; Neutrino 2012 conference, http://neu2012.kek.jp/.

[3] Cabibbo N 1963 *Phys. Rev. Lett.* **10** 531; Kobayashi M and Maskawa T 1973 *Prog. Theor. Phys.* **49** 652; Nakamura K et al. 2010 Particle Data Group *J. Phys. G: Nucl. Partl. Phys.* **37** No.7A

[4] Pontecorvo B 1968 *Sov. Phys. JETP* **26** 984; Maki Z, Nakagawa M and Sakata S 1962 *Prog. Theor. Phys.* **28** 870

[5] King S F and Singh N Nimai 2000 *Nucl. Phys.* B **591** 3; 2001 *Nucl. Phys.* B **596** 81

[6] Pascoli S and Petcov S T 2008 *Phys.Rev.* D **77** 113003

[7] Thomas Shaun A et al. 2010 *Phys. Rev. Lett.* **105** 031301

[8] Tzanakos George S(Ed.) 2012 NEUTRINO 2010 *Nucl. Phys.* B (Proc. Suppl.)

[9] Gell-Mann M, Ramond P and Slansky R 1979 *Supergravity, Proc. Workshop* (New York: Stony Brook), (Eds) Nieumenhuizen P Van and Freedman D 1979 (Amsterdam: North Holland); Yanagida T 1979 *KEK Lectures* (unpublished); Mohapatra R N and Senjanovic G 1980 *Phys. Rev. Lett.* **44** 912

[10] Harrison P F and Scott W G 2002 *Phys.Lett.* B **547** 219; Harrison P F, Perkins D H and Scott W G 2002 *Phys. Lett.* B **530**167

[11] Altarelli G and Feruglio F 1999 *Phys. Rep.* **320** 295; Patgiri M and Singh N Nimai 2003 *Phys. Lett.* B **567** 69

[12] Harrison P F, Perkins D H and Scott W G 2002 *Phys. Lett.* B **530** 167; Singh N Nimai, Rajkhowa M and Borah A 2007 *J. Phys. G: Nucl. Partl. Phys.* **34** 345

[13] Duarah C and Singh N Nimai 2012 *Phys. Lett.* B **718** 147

[14] King S F 2012 *Phys. Lett.* B **718** 138

[15] Fritzsch H and Xing Zhi-zhong 1998 *Phys. Rev.* D **57** 594

[16] Duarah C and Singh N Nimai 2013 *Indian Journal of Physics* (IJP) (to appear); *HORIZON-A Journal of Physics* 2012 **2** 73

[17] Roy S and Singh N Nimai 2012 *Horizon -A Journal of Physics* **2** 65 (*Preprint* 1211.7207 [hep-ph])

[18] Wilczek F 2009 *Nature Physics* **5** 614; Singh N Nimai 2011 *Horizon-A Journal of Physics* **1** 60

[19] Francis Ng K and Singh N Nimai 2012 *Nucl. Phys.* B **86** 19

[20] Lambiase G, Mishra H and Mohanty S (*Preprint* 1006.4461 [hep-ph]); Bjaelde O E and Das S (*Preprint* 1002.1306 [hep-ph])

[21] Bhatt J R, Desai B R, Ma E, Rajasekaran G and Sarma U 2010 *Phys. Lett.* B **687** 75; Bhatt J R and Sarma U *Phys. Rev.* D **80** 045016

Is the scalar Higgs boson observed at the LHC, a supersymmetric Higgs boson or a standard model one?

Subrata Chakraborty

Department of Physics, Visva Bharati University, Santiniketan 731235, India.

E-mail: subrata502002@yahoo.co.in

Abstract. Recently, a SM-like Higgs particle is reported to be discovered at CMS and ATLAS experiments at CERN with a mass of about 124 GeV. The present paper is related to an attempt to understand a very important question: Is the scalar Higgs boson observed at the LHC, a supersymmetric Higgs boson or a standard model one? Actually, after briefly reviewing the theoretical studies on Higgs mass in the context of Standard Model (SM) and Minimal Supersymmetric Standard Model (MSSM), we compare the values of Higgs mass obtained in several theoretical works on SM and MSSM, including those of the present author. We comment critically as regards the identification of the Higgs boson as SM Higgs or MSSM Higgs. In particular, certain models show that MSSM can give the mass of the lightest Higgs at a value near the observed value of nearly 124 GeV of SM-like Higgs, for a suitable choice of parameters. We suggest that repeated experimental confirmation as well as studies of other distinguishing properties of the discovered Higgs particle can probably discriminate the SM Higgs boson from the lightest MSSM one?

1. Introduction

Recently a Higgs-like particle is discovered in Large Hadron Collider experiment [1] at CERN on July 4,2012 and this first LHC result suggests that the Higgs mass lies at 125-126 GeV. In fact,the standard model is a very successful model of strong and electroweak interaction and is due to the independent works of Glashow, Weinberg and Salam [2]. All predictions of the model has met with extreme success except that the Higgs particle, required for giving mass to the particles of the standard model other than the photon, is not yet confirmed experimentally. This particle was predicted by Higgs and other scientists [3]. Unfortunately, the Higgs mass is a free parameter in the standard model.

In particle physics, symmetries are believed to permit ultimately a classification of all observed particles. Supersymmetry (SUSY) arises as a symmetry [4] which combines bosons and fermions in the same representation or multiplet of the enlarged group which encompasses both the transformation of the Poincare group and the appropriate transformations, so that every bosonic particle must have a supersymmetric partner and vice versa. Since such a spectrum of particles is not compatible with observations, SUSY must be broken badly at the level of presently available energies. Experimental observation of a Higgs particle of supersymmetry,for example,can help us decide as to whether supersymmetry indeed is inherent in nature.

The purpose of the present paper is to throw light on the identification of the discovered Higgs on the basis of several theoretical studies on SM and MSSM Higgs i. e. whether theory favours it to be SM Higgs or a SUSY Higgs boson. Interpretation of the new particle within SM or within supersymmetric Higgs sector and conclusions about the nature of the Higgs boson (SM or SUSY etc.) are necessary.

The paper is organised as follows: a short introduction to a few theoretical models of SUSY Higgs and the Higgs mass predictions are given in section 2; the problem of discrimination among SM Higgs and MSSM Higgs are analysed in section 3. Discussions are given in section 4,where the role of symmetry, aesthetics and beauty in the theoretical high energy physics are also mentioned. Conclusions are given in section 5.

2. A brief review of theories and theoretical predictions on SUSY Higgs mass

Supersymmetry(SUSY) is one of the most beautiful theoretical ideas that tries to explain the origin of the scale of the electroweak interaction. The minimal supersymmetric standard model(MSSM) is the most economical models and gives physics beyond the standard model(SM). In MSSM every quark,lepton and gauge boson has a supersymmetric partner associated with it. Moreover,the MSSM requires two Higgs doublets so as to give masses to up-type and down-type fermions, by keeping consistency with supersymmetry and by avoiding gauge anomalies due to the fermionic super-partners of the Higgs bosons. In MSSM ,after electroweak symmetry breaking there remain a total of five physical Higgs bosons. These can be classified as two neutral CP-even bosons,one neutral CP-odd boson and two charged bosons. The formalism of MSSM and mathematical details of calculations of the tree-level masses of supersymmetric Higgs may be found in a review [5], where the five tree level Higgs masses (the symbols having their usual meaning) satisfy the bounds given as follows

$$m_{H^\pm} \geq M_W, \tag{1}$$

$$m_{H_2^0} \leq m_Z \leq m_{H_1^0}, \tag{2}$$

$$m_{H_3^0} \geq m_{H_2^0}. \tag{3}$$

Although at tree level the condition $m_{H_2^0} \leq m_Z$ holds, no Higgs has ever been detected at LEP and hence there is necessity of one and two-loop radiative corrections. For models which give 1-loop radiative corrections we refer to the work [6]. For some related works giving the lightest Higgs mass in MSSM to be $< 130GeV$ one may see [7]. A recent work for Higgs mass in MSSM is given in [8]. For detailed studies on phenomenological aspects of SUSY particle one may see [9].

A compilation of Higgs mass predictions in the standard model and supersymmetric standard model is given in the work of ref. [10]. Out of a list containing 64 Higgs mass predictions mentioned here, supersymmetry is behind 18 of them and with the exception of two,their central values lie between 120 and 160 GeV. Some of these predictions compiled in ref. [10],which lie near the observed value of Higgs mass (125 GeV), are shown in table1.

3. The problem of discrimination among SM Higgs and MSSM Higgs boson

In an article Jason L. Evans [11] has stated that SUSY prefers light Higgs boson but SM does not care; post LEP bounds are 114 - 600 GeV while SUSY expected mass range is 90 to 130 GeV. This author has posed the question : if the Higgs particle discovered is SM Higgs, why did it fall in the MSSM window?In another article Pran Nath [12] has assumed SUSY Higgs to be the underlying model for explaining the new boson by considering SUSY breaking. He concluded that the Higgs boson mass is less than 150 GeV (130 GeV) in MSSM (mSUGRA). He also concluded that the prospect for SUSY discovery at LHC are brighter as a consequence of the discovery of the Higgs boson. In another article entitled Consequences of a 125 GeV

Journal of Physics: Conference Series **481** (2014) 012003 doi:10.1088/1742-6596/481/1/012003

Table 1. A compilation of Higgs mass (m_H(GeV)) predictions [10],which are near the recent LHC experimental result (125 GeV) are shown here

m_H (GeV)	Authors	Idea	Techniques
120 ± 6	Ellis et al (2005)	Supersymmetry	MSSM
121 ± 6	Feldman et al (2006)	Superstring inspired	RG flow
121.8 ± 11	Froggatt et al (1995)	Degenerate vacua	RGE
122 ± 10	Djouadi (2004)	SUSY SU(5)	RG flow
124 ± 21	Barger et al (2007)	Susy broken	R G flow
124 ± 10	Arbuzov et al (2007)	Condensates etc.	Coleman-Weinberg
124.2 ± 13.2	Codoban et al (1999)	Supersymmetry	MSSM
125 ± 4	Gogoladze et al (2007)	Extra-dimension	R G flow
127.5 ± 7.5	Chankowski et al (2004)	Susy	MSSM

Table 2. Maximum value of lightest Higgs mass M_h in different models of MSSM [13]

M_h^{max}	mSUGRA	no scale	CNMSSM	VCMSSM
value	128.0	123.0	123.5	124.5

Higgs on constrained MSSM scenarios by Mahmoudi [13] the author suggests that SM gives a Higgs mass which is essentially a free parameter, whereas MSSM gives the lightest CP-even Higgs particle as bounded from above. In different models of MSSM ,maximum value of lightest MSSM Higgs mass M_h lies near 125 GeV approximately (see table2 [13]). The main purpose of this section has been to mention about the theoretical results on supersymmetric Higgs related to the LHC result. Probably SUSY could be discovered even before the Higgs [14]! The current SUSY Higgs particle mass bounds are given elsewhere [15].

4. Discussions

The present studies are based on predictions of theoretical models of supersymmetry on Higgs mass by different authors including ours. The Higgs boson discovered is also likely to be MSSM Higgs based on the predictions compiled in tables 1 and 2 , although one cannot surely distinguish between SM and MSSM Higgs at this stage.

In the context of LHC result on Higgs mass, we try to pin point the problem by going into details of theories on SM and MSSM. May be the Higgs particle is lightest SUSY Higgs particle, if discrepancy is noticed with respect to SM. The problem of resolving the issue of newly discovered boson may be understood through independent discovery,with caution in accepting new ideas and after critical enquiry into the evidence of discovery of the Higgs particle.

Hence connecting the theoretical models, bounds and constraining the parameters are very important. We may use symmetry to choose the correct option for the theory. Often,the symmetry principles [16] are the most important guides for understanding the unknown domain of nature. The concept of gauge invariance,formal relations between symmetries, unification and conservation laws are found to be very useful. Symmetry should be the guiding principle in the analysis of experimental data with reference to the theoretical models in high energy physics. There is a role of aesthetics in scientific theories [17]. One may use aesthetics in science to choose the most beautiful option in the matter of introduction of new ideas through economy, simplicity, harmony and unified description giving a connection. The theory should be based on unbiased experimental results [18]. Beauty in particle physics [19] should also be considered.

5. Conclusions

The experimentalist, observation on Higgs-like particle and the analyses of theoretical results on models of SM and MSSM are likely to give feedback to the future search of the particle. Confining our attention to the mass spectra,we give a general survey based on theoretical works of different workers including ours. The MSSM Higgs mass varies with different parameters which are not precisely known as yet. We suggest that symmetry, aesthetics and match with experimental data are to be used for selecting the correct theory in SM and MSSM. If SM-like Higgs is SM Higgs, why does the Higgs mass predicted by many SUSY models lie in the range of values near the LHC value of Higgs Mass?

In this paper we have reviewed the current situation in LHC and Higgs mass to some extent. We think that more exhaustive and systematic studies on both SM and MSSM are necessary. We comment on the value of Higgs mass in LHC vs. predictions of Higgs mass in SM and MSSM [10]. We tabulated the predictions by different workers. On the basis of experimental data in LHC and the present theoretical scenario we come to the conclusion that the Higgs particle discovered is likely to be SM-like MSSM lightest Higgs. Sometimes it may happen that scientists trying to discover one thing fail to do so but they may discover some other new thing by chance. Unfortunately,the Higgs mass bounds in SM is over a large range (114 - 600 GeV). Probably, SUSY is discovered through the LHC experiment in July 2012. Its a great triumph for the experimentalists(CMS and ATLAS in CERN) as well as the theorists who have predicted the Higgs mass in SUSY over this range. It is hoped that more experiments are needed in order to confirm or reject this viewpoint. This study also reveals that there is a confusion regarding the true identity of the particle discovered in LHC experiments in July,2012. On the basis of a critical analysis we come to the following conclusions

- undoubtedly the experiment should be repeated at least two more times and the result should be checked
- the detail properties of the discovered (Higgs!) particle should be found out
- once it is firmly established that it is a Higgs particle,the problem remains as to whether it is a SM Higgs, MSSM-Higgs or a new particle
- since supersymmetry is a more complete theory than the standard model and the mass of the lightest Higgs in SUSY is more probable in view of the recent discovery at CERN, there may be a probability that the discovered particle is a Higgs particle of the Supersymmetric Standard Model.

Hence these ideas are useful for successfully connecting the theoretical models and bounds on Higgs mass and constraining the parameters for a true theory compatible with confirmed experimental result. Only those theoretical results matching with experimental data may be given importance,which are based on general principles of symmetry, aesthetics, beauty and scientific methodology of research. The Higgs like particle discovered at CERN in 2012 may be a SM Higgs or a MSSM lightest Higgs. More experiments about the detailed properties of the particle are needed in order to confirm these aspects. All these new experimental results should be reproduced before final conclusion.

6. Acknowledgments

The author would like to express heart felt thanks to Profs N Nimai Singh, Kalpana Bora, Madhurjya P Bora and other members of the Organising Committee for invitation, partial financial support and excellent hospitality. I would also like to thank the Organising Committee members for giving me an opportunity to play 'Bihu' tune on Violin in the Cultural Programme arranged for the participants of the conference.

References

[1] Large Hadron Collider - Wikipedia,the free encyclopaedia*en. wikipedia. org/wiki/Large-Hadron- Higgs-like Particle Discovered at CERN*; Kane G, Pierce A (ed.)2008*Perspectives on LHC physics* (Singapore:World Scientific)

[2] Glashow S L 1961 *Nucl. Phys.* **22** 579; Weinberg S 1967 *Phys. Rev. Lett.* **19** 1264 ; Salam A 1968 it Elementary Particle Theory (Stockholm:Almquist and Forlag)

[3] Higgs P W 1964*Phys. Lett.* **12** 132; *Phys. Rev. Lett.* **13** 508; Englert F and Brout R 1964 *Phys. Rev. Lett.* **13**, 321; Guralnik G S, Hagen C R and Kibble T W B 1964 *Phys. Rev. Lett.* **13**585

[4] Kim J 1991 *The Standard Model and Beyond* (Singapore:World Scientific)p 304; Nilles H P 1984 *Phys. Rep.* **110** 1;Haber H and Kane G 1985 *Phys. Rep.* **117** 75; Gunion J F,Haber H E, Kane G L and Dawson S 1990 *The Higgs Hunter,s Guide* (Redwood City,CA :Addison-Wesley) erratum,1992 Report No. SCIPP-2/58(unpublished); Barbieri R 1988 *Riv. Nuovo Cimento* **11** 1

[5] Simonsen I 1995 A Review of Minimal Supersymmetric Electro Weak Theory *Preprint* hep-ph/9506369

[6] Haber H E,Hempling R and Hoang A H 1997*Z. Phys. C.* **75** 539 ; Haber H E and Hempfling R 1993 *Phys. Rev.* **D 48**4280 and references there in.

[7] Ghosh B and Chakrabarty S 2001 *Ind. J. Phys.* **75A(6)** 651; Ghosh B and Chakrabarty S 2001 *Ind. Jour. Pure and Applied Physics* **39** 481

[8] Chakraborty S and Pal P P 2010 *Journal of Assam Science Society* **51** No 1 and 2

[9] Muller-Kirsten H J W and Wiedemann A 1987 *Supersymmetry: An Introduction with Conceptual and Calculational Details*(Singapore: World Scientific); Drees M, Godbole R M and Roy P 2004 *Theory and Phenomenology of Sparticles* (Singapore:World Scientific);Baer H A and Tata X R 2006 *Weak Scale Supersymmetry: From Superfields to Scattering Events* (Cambridge, England:Cambridge University Press)

[10] Schucker T Higgs-mass predictions,Centre de Physique Theorique *CNRS preprint* No. 0708. 3344

[11] Evans J L 2001 *Phys. Rev.* **D 86** 015017; Muhlleiner M M Invited review in *Int. J. Mod Phys A***27**12300 30 20th International Conference on Supersymmetry and Unification of Fundamental Interactions (SUSY 2012); Carena M 2012 *Higgs Boson Discovery and Supersymmetry at the LHC*SUSY 2012

[12] Nath P 2012 *Perspectives on Higgs and Supersymmetry,SUSY 2012* (Beijing,China)

[13] Mahmoudi F 2012 *Higgs Physics and SM/BSM in the light of LHC data* SUSY 2012

[14] Djouadi A 2008 *Phys. Rep.* **457** p 1-216 ;**459** p 1-241

[15] Haber H E 2002 *Higgs Theory and Phenomenology in the Standard Model and MSSM arxiv. org /pdf/hep-ph* /0212136; Carena M and Haber H E 2003 *Prog. Part. Nucl. Phys* **50**63-152 *arXiv:hep-ph/*0208209 ; Das S P and Drees M 2011 *Phys. Rev.* **D 83** 035003 and references there in

[16] Khanna M P 1999*Introduction to Particle Physics* (New Delhi:Prentice-Hall of India)Ch. 3, p. 70; Symmetry and Symmetry Breaking, Dec 2, 2004 (Stanford Encyclopedia of Philosophy) *http://plato. stanford. edu/entries/symmetry-breaking/*

[17] Chandrasekhar S 1987 *Truth and Beauty: Aesthetics and Motivations in Science* (Chicago,Illinois: The Chicago University Press)

[18] Feynman R P 1970 *The Character of Physical Law*(Boston:The M. I. T. PRESS, Third Paperback Printing); Gregory Sir R 1949 *Discovery* (Puffin Story Books, Penguin Books); Osheroff D D 2001 *Am. J. Phys.* **69** 26 ;Kothari C R 1998 *Research Methodology (Methods and Techniques)*(New Delhi:Wishwa Prakashan)

[19] Chakrabarty S 1996 *Beauty in Elementary Particle Physics: Aesthetics and Motivations in Arts and Science* Ed. Gupta K C (New Delhi:New Age International Limited) Chap. 13, p. 173-178

Dark energy model building and observational issues

Anjan A Sen
Center for Theoretical Physics, Jamia Millia Islamia, New Delhi 110025, India
E-mail: aasen@jmi.ac.in

Abstract. In this short review, I shall discuss the issues relating to dark energy and late time acceleration of the Universe. I shall give a brief overview of the possible model building approaches for the late time acceleration of the Universe as well as different observational issues regarding dark energy.

1. Introduction

The most pathbreaking discovery in cosmology in recent years is the observation of cosmological acceleration in our Universe. Starting in the year 1999 when two independent observational efforts relating to the Type-Ia supernova first pointed out that the Universe may be accelerating recently, it has now been confirmed by a variety of cosmological observations [1]. The big picture tha has now emerged from a host of very accurate cosmological observations are as follows: we are living in a Universe which is homogeneous and isotropic on large scales. This Universe is well described by a flat FRW Universe on large scales. The visible part of the Universe constitutes only 5 percent of the total energy budget of the Universe. The rest of the Universe is dark with 25 percent constitutes with non-relativistic dark matter, while the rest 70 percent of the Universe is given by a smooth relativistic component having a negative pressure whose magnitude is comparable to its rest energy density. This component with negative pressure is called the dark energy which is responsible for the accelerating Universe.

To understand the effect of dark energy in simple terms, let us write the equaton of motion in Newtonian Gravity. It is given by

$$\frac{\ddot{R}}{R} = -4\pi G\rho, \tag{1}$$

Here ρ is the rest energy density of the gravitating matter. As this is always positive, the negative sign in the rhs of the above equation (which is due to the attractive nature of gravity) ensures that we have always deceleration. But when we write the same equation in Einstein General Relativity (GR), this equation gets a crucial modification. Due to the fact that pressure also gravitates in Einstein GR, this equation becomes:

$$\frac{\ddot{R}}{R} = -4\pi G(\rho + 3p), \tag{2}$$

where we assume that space part of the Universe is homogeneous and isotropic and p is the pressure in all three directions. The remarkable property of the equation is that if $\rho + 3p) < 0$

Journal of Physics: Conference Series **481** (2014) 012004 doi:10.1088/1742-6596/481/1/012004

(also known as violation of strong energy condition), then $\ddot{R} > 0$ and the Universe will accelerate. Hence to accelerate the Universe, we need a dark component which has negative pressure such that $\rho + 3p < 0$.

Let us see the possible ways one can explain the late time acceleration of the Universe. The Einstein equation is given by

$$G_{\mu\nu} = 8\pi G T_{\mu\nu} \tag{3}$$

With this if we assume a FRW metric to describe our Universe in cosmological scales, there is essentially two ways to explain the late time acceleration. If we concentrate on the rhs of the Einstein equation, we have already explained above that one needs to have negative pressure for acceleration. And as no normal component that we know has negative pressure, one has to include some additional exotic component having a negative pressure (we call it Dark Energy) to explain the late time acceleration. On the other hand if we keep normal components in the rhs of the Einstein equation, then one has to modify the lhs of the equation in such a way so that any departure from simple $G_{\mu\nu}$ term will mimick the negative pressure component that we require to add in the rhs of the equation to explain the acceleration [2].

There is a third way out. In this case, one retains the Einstein equation as it is, also keep the normal component in the rhs, but give up the idea that Universe is described by a FRW metric. In this review I shall only describe the first approach that is to say that we need to add some exotic component with negative pressure in the energy budget of the Universe to explain the late time acceleration.

Now how to probe this mysterious dark energy? In simple terms, there are three possible ways. One is to probe its effect on background cosmological expansion. This includes geometric quantities like distances, angles, volumes etc. The observables for this are standard candles like Sne-Ia, standard rulers like Acoustic Oscillations in Baryon-Photon fluid before decoupling. The second one is the direct or indirect effect of dark energy on structure formation of the Universe. The observables in this case are linear and non linear clustering, the Integrated Sachs-Wolf effect in temperature anisotropy in CMB, redshift space distortion, weak lensing etc.

2. Dark Energy Models

Simplest example for dark energy is a cosmological constant which is described by the equation of state $p = -\rho$. The cosmological constant together with the normal non relativistic matter (known as the concordance ΛCDM model) describes perfectly all the cosmological observations till date. But this is plagued by problem of unnatural fine tuning as well as the cosmic coincidence problem [1]. Also the observational data allow models different from ΛCDM where at least the cosmic coincidence problem can be addressed successfully.

One of the alternatives to ΛCDM model is canonical scalar field slowly rolling over a sufficiently flat potential at present epoch [3]. The idea is same as that of a inflation field; the only difference is that the energy scale involved is much lower (10^{-3} eV). The equation of motion for a scalar field in an expanding FRW background is given by

$$\ddot{\phi} + 3H\dot{\phi} + \frac{\partial V}{\partial \phi} = 0. \tag{4}$$

The evolution of the scalar field is driven by the steepness of the potential and it is opposed by the friction term due to the expanding background. Depending upon how these two terms interplay, one can broadly classify the scalar fields into two categories: the freezing models and the thawing models [4]. For the freezing models, the scalar fields initially has a fast roll phase due to a steep potential and this helps it to mimick the background Universe. One can fine tune the form of the potential in such a way that around the present epoch the scalar field starts slow rolling and can give rise to negative pressure. On the other hand, thawing models have sufficinetly flat potentials and is frozen to $w = -1$ initially due to large Hubble friction initially. As the Universe evolves, the Hubble friction decreases, and the scalar field slowly thaws away from its frozen $w = -1$ state. In this model, the equation of state of the scalar field is always very close to $w = -1$ and hence it behaves as dark energy without any fine tuning of the form of the potential, but one has to be fine tune the initial condition to get the right amount of scalar field energy density at present. One can use the different observational data to calculate the Bayesian Evidence for both thawing and freezing class of models [5]. The Bayesian Evidence is given by

$$E = \int P(\theta)\mathcal{L}(\theta)d\theta \tag{5}$$

where θ's are the model parameters, $P(\theta)$ is the prior probability distribution for the parameters θ's and $\mathcal{L}$ is the likelihood function. If one considers the dark energy to be homogeneous, there is a significant evidence in favour of the thawing class of models. But assuming the dark energy to be clustered, both thawing and freezing class of models are equally preferres by the observational data [5].

One can consider more exotic fields to be considered as candidate for dark energy. One such example is the Galileon field given by the lagrangian (See [6] and references therein):

$$\mathcal{S} = \int d^4x\sqrt{-g}\left[\frac{M_p^2}{2}R - \frac{1}{2}(\nabla\pi)^2\left(1 + \frac{\alpha}{M^3}\right) - V(\pi)\right] + \mathcal{S}_m \tag{6}$$

where α is strength of the higher dimensional operator that is present in the action and M is fifth dimensional Planck's mass. The good part of the this action is that despite having higher derivative term in the action, the equation of motion is still second order. Also the field π respects the Galileon symmetry in the Minkowski background without the potential or with a linear potential.

One can now calculate the Bayesian Evidence for this model for different potentials and compare them with the standard scalar field with canonical kinetic term (without any higher derivatives in the action) using different observational data. It has been shown that for linear and quadratic potential there are very strong evidence for this Galileon model in comparison to standard scalar field, whereas for exponential and inverse quadratic potential both are equally favored [6].

3. Conclusion

To conclude, the fact that our Universe is currently going through an accelerating expanding phase is now confirmed by various observations. Although a simple cosmological constant is most suited model for all the observations, the theoretical problems associated with such a set up, motivates people to look for the other options. Scalar field models are the simplest alternatives. With present observational results, one cannot distinguish various kind of scalar field models. But if one goes beyond the simple scalar fields and consider more exotics actions for these fields, present data can indeed distinguish them for some specific potentials.

Acknowledgements

I thank the organizers of the NCHEPC-2013 for inviting me to give this talk.

References

[1] Miao Li, Xiao-Dong Li, Shuang Wang, arXiv:1103.5870; V. Sahni and A. A. Starobinsky, Int. J. Mod. Phys. D **9**, 373 (2000); S. M. Carroll, Living Rev. Rel. **4**, 1 (2001); P. J. E. Peebles and B. Ratra, Rev. Mod. Phys. **75**, 559 (2003); T. Padmanabhan, Phys. Rept. **380**, 235 (2003);
[2] S. Tsujikawa, e-Print: arXiv:1101.0191 [gr-qc].

[3] E. J. Copeland, M. Sami and S. Tsujikawa, Int. J. Mod. Phys. D **15**, 1753 (2006)
[4] Caldwell R. R. & Linder E. V., Phys. Rev. Lett. **95**, 141301 (2005).
[5] S. Thakur, A. Nautiyal, A. A. Sen and T. R. Seshadri, MNRAS, **427**, 988, (2012).
[6] M. W. Hossain and A. .A. Sen, Phys. Lett. B., **713**, 140, (2012).

Journal of Physics: Conference Series **481** (2014) 012005 doi:10.1088/1742-6596/481/1/012005

Bounds on sterile neutrino parameters from reactor experiments[1]

Kalpana Bora

Physics Department, Gauhati University, Guwahati 781014, Assam, India.

E-mail: kalpana@gauhati.ac.in

Abstract. In this work, we present a realistic analysis of the potential of the present-day reactor experiments Double Chooz, Daya Bay and RENO for probing the existence of sterile neutrinos. We present exclusion regions for sterile oscillation parameters and find that these experimental set-ups give significant bounds on the parameter Θ_{ee} especially in the low sterile oscillation region $0.01 < \Delta m^2_{41} < 0.05$ eV^2. These bounds can add to our understanding of the sterile neutrino sector since there is still a tension in the allowed regions from different experiments for sterile parameters.

1. Introduction

A few years back, experiments like LSND [1] and MiniBooNE [2] have indicated the presence of a fourth type of neutrino - the sterile neutrino. The LSND result provided evidence for $\bar{\nu_\mu} \to \bar{\nu_e}$ oscillations from μ^+ decay-at-rest(DAR), and the same oscillation channel was studied in the MiniBooNE experiment. Similar results are also indicated by $\bar{\nu_e}$ and ν_e disappearance channels, by the reactor anomaly [3] and Gallium anomaly [4,5], and from cosmology [6,7]. These results suggest that the deficit of observed neutrino fluxes in the respective channels (ν_e or $\bar{\nu_e}$) may be an indication of the existence of a fourth type of neutrino. The latest global fits of sterile neutrino parameters were presented in [8,9].

A significant amount of work is already available in the literature regarding the search for sterile neutrinos in reactor and atmospheric neutrino oscillation experiments. Recently the possibility of using atmospheric neutrinos as a probe of eV^2-scale active-sterile oscillations was explored in [10], where bounds on $\sin^2 2\Theta_{\mu\mu}$ and Δm^2_{41} were presented. The implication of sterile neutrinos on measurements of θ_{13} in a $(3+2)$ scenario in the Double Chooz [11] reactor experiment was studied in [12]. The impact of light sterile neutrinos on θ_{13} measurements in Double Chooz and Daya Bay [13] was studied in [14] in a (3+1) scenario. A study of the effect of sterile neutrinos on θ_{23} and θ_{13} measurements in MINOS, T2K and Double Chooz was performed in [15]. Similar studies for Daya Bay were carried out in [16]. A search for sterile neutrinos using MINOS was done in [17]. A constraint on the mixing angle θ_{14} from a combination of Solar and KamLAND data was given in [18]. An analysis of the results of Double Chooz, Daya Bay and RENO to simultaneously fit θ_{13} and the reactor neutrino anomaly was recently performed in [19].

[1] Based on: K. Bora, D. Dutta, P. Ghoshal, JHEP 12 (2012) 025; arXiv:1206.2172v2 [hep-ph]

In this work, we present exclusion regions in the sterile neutrino parameter space $\sin^2 2\Theta_{ee} - \Delta m^2_{41}$ for the three current reactor experiments, namely Double Chooz, RENO [20] and Daya Bay. A similar study was performed in [14] for Double Chooz and Daya Bay, where a more approximate analysis in the $\sin^2 2\Theta_{ee} - \Delta m^2_{41}$ plane was done assuming an overall systematic error and neglecting detector resolution. In our present work, we have used simulations with reduced values of errors, as quoted in the technical reports of the individual reactor experiments, where the cancellation of correlated reactor-related errors by using both near and far detectors has been taken into account. Also, we have considered realistic detector resolutions, which play an important part in the sensitivity analysis. . We also compare our results with those of older reactor experiments like BUGEY [21], GOSGEN [22] and Krasnoyarsk [23].

2. Neutrino mixing-$(3+1)$ scenario

In presence of a sterile neutrino, three flavor Pontecorvo-Maki-Nakagawa-Sakata (PMNS) neutrino mixing changes to (3+1) or (2+2) scenario. Here we consider the (3+1) scenario, which is an extension of the three-neutrino scenario with the addition of one massive sterile neutrino. Solar and atmospheric neutrino analyses strongly discourage the (2+2) scenario due to the absence of sterile signals in the atmospheric parameters [24]. The (3+2) scenario is favoured by the tension between appearance and disappearance experiments in the (3+1) case, but disfavoured by cosmology.

In the presence of a sterile neutrino, the standard 3-flavour neutrino oscillation picture changes, and hence the survival probability must be modified due to the effect of (3+1) mixing. The baselines relevant to these experiments are of the order of a few hundred metres, and hence the oscillation would show signatures at the atmospheric scale as well as possible sterile-scale effects. The electron neutrino survival probability expression relevant for our analysis is [14]

$$P_{ee} = 1 - \cos^4\theta_{14}\sin^2 2\theta_{13}\sin^2(\frac{\Delta m^2_{13}L}{4E}) - \sin^2 2\theta_{14}\sin^2(\frac{\Delta m^2_{14}L}{4E}), \quad (1)$$

where L is the baseline and E is the neutrino energy. Here, P_{ee} is seen to be a function of two mass-squared differences (Δm^2_{14} and Δm^2_{13}) and two mixing angles θ_{13} and θ_{14}. When $\theta_{14} \to 0$, we recover the standard three-flavour probability expression in which the solar mass scale is neglected.

3. Technical details of the experiments

In this section, we present some technical details of the three new generation reactor experiments, Double Chooz [11, 20], Daya Bay [13] and RENO [20], for the sake of completeness, collected from [11, 13, 20]. All the information regarding these experiments are given in table 1. The systematic errors associated with each experimental set-up are listed in Table 2.

The details of the three old reactor experiments (BUGEY [21], Gosgen [22] and Krasnoyarsk [23]) with which we compare our exclusion regions for sterile parameters are tabulated in Table 3.

4. Results and analysis

We have generated the 90% c.l. exclusion plots for sterile oscillations for all the three current experiments and compared the results with the old reactor experiments Bugey, Gosgen and Krasnoyarsk. The right side of each contour shows the no-oscillation region while the region left to the contours is the possible oscillation region. The results are found to be very sensitive to the values of systematic errors. In our calculations, we have used GLoBES [25–27] for simulating

Journal of Physics: Conference Series **481** (2014) 012005 doi:10.1088/1742-6596/481/1/012005

Table 1. Details of the three reactor experiments

Name of Exp	Double Chooz	Daya Bay	RENO
Location	France	China	Korea
No of Reactor cores	2	4	6
Total Power(GW_{th})	8.7	11.6	16.4
Baselines- near/Far(m)	410/1067	363(481)/1985(1615)	292/1380
Target mass(tons)	10/10	$40 \times 2/10$	16.1/16.1
No of Detectors	2	2	2
Exposure(years)	3	3	3
Resolution(%)	12	12	12

Table 2. Systematic errors associated with the three experiments

Name of Exp	Double Chooz	Daya Bay	RENO
Reactor correlated error(%)	2.0	2.0	2.0
Detector normalisation error(%)	0.6	0.5	0.5
Scaling or calibration error(%)	0.5	0.5	0.1
Overall normalization error(%)	2.5	2.5	2.0
Isotopic abundance error(%)	2.0	2.0	2.0

Table 3. Details of old reactor experiments

Name of Exp	Bugey	Gosgen	Krasnoyarsk
No of Reactor cores	4	1	3
Total Power(GW_{th})	2.8	2.8	2.8
Baselines(m)	15,40,95	37.9,45.9,64.7	57,57.6,231.4
Target mass(≈tons)	1.67	0.32	0.4
No of Detectors	1	1	1
Exposure(≈ years)	0.2	0.39,0.56,0.98	0.06
Resolution(%)	6	-	-

the experiments. In our calculation we have used the χ^2 function as:

$$\chi^2 = \sum_{i=1}^{\#\text{of bins}} \sum_{d=N,F} \frac{(O_{d,i} - (1 + a_R + a_d)T_{d,i})^2}{O_{d,i}} + \frac{a_R^2}{\sigma_R^2} + \frac{a_N^2}{\sigma_N^2} + \frac{a_F^2}{\sigma_F^2} \tag{2}$$

where $O_{N,i}, O_{F,i}$ are the event rates for the ith bin in the near and far detectors, calculated for true values of oscillation parameters; $T_{d,i}$ are the expected event rates for the ith bin in the near and far detector for the test parameter values; a_R, a_F, a_N are the uncertainties associated with the reactor flux and detector mass; and $\sigma_R, \sigma_F, \sigma_N$ are the respective associated standard deviations.

Fig.1 shows the variation of the average bound on $\sin^2 2\theta_{ee}$ from each of the three experiments Double Chooz, Daya Bay and RENO as a function of the bin-to-bin systematic error for a

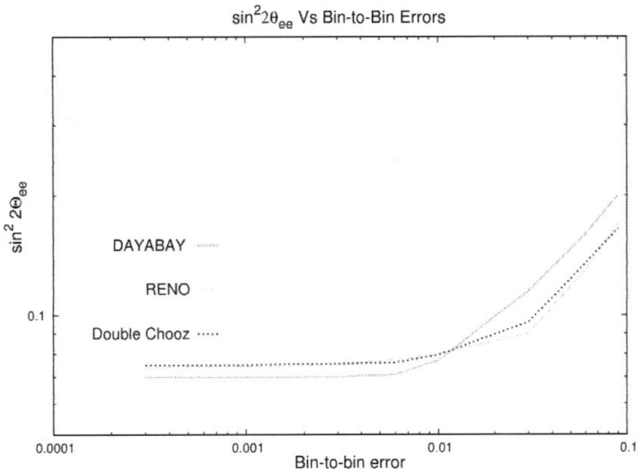

Figure 1. Dependence of the average $\sin^2 2\theta_{ee}$ bound on the correlated bin-to-bin error for Double Chooz, Daya Bay and RENO.

Table 4. Set of reduced errors used in our calculation

Name of Exp	Double Chooz	Day Bay	RENO
Reactor correlated error(%)	0.06	0.087	0.5
Detector normalisation error(%)	0.06	0.12	0.5
Scaling or calibration error(%)	0.5	0.5	0.1
Overall normalization error(%)	0.5	0.5	0.5
Isotopic abundance error(%)	0.06	0.087	0.5

constant overall normalisation. We depict the behaviour with respect to the correlated bin-to-bin error since this is found to have the most significant effect on parameter sensitivities. The figure shows that the dependence of the sensitivity in the case of Daya Bay is slightly steeper, even for low values of the bin-to-bin systematics, than for the other two experiments.

In our further analysis, we have used a reduced set of systematic errors as listed in Table 4, taking into account the partial cancellation of errors due to the presence of both near and far detectors, as documented in the experiment literature.

We have included the changed flux normalisation given by the reactor flux anomaly [3] in our sensitivity analysis. We are performing a simulation and not a data analysis, and hence varying the flux normalisation has minimal effect on our results, since the relative normalisation simultaneously affect both the true and test spectra in the simulation and lead to a cancellation of their effect in the parameter sensitivity. For the same reason, leaving the flux normalisation as a freely varying parameter does not have a major influence on our results, and therefore they may be taken to be indicative of the effect of spectral information only.

In Fig.2, we show the comparison of the exclusion plots for Double Chooz, RENO and daya Bay experiments with existing reactor experiments using modified errors. Our results for all the three experiments show better exclusion regions for sterile neutrino oscillation parameters than Gosgen and Krasnoyarsk in the range Δm^2_{41}=0.01 to 1 eV^2. Our bounds are an improvement over Bugey in the range Δm^2_{41}=0.01 to 0.05 eV^2 but in the Δm^2_{41} region above this, Bugey gives better bounds. From these curves, we see that $\sin^2 2\theta_{ee} > 0.1$ and $\sin^2 2\theta_{ee} > 0.08$ is excluded for sterile oscillation in the Δm^2_{41}=0.01 to 1eV^2 region for Double Chooz and RENO respectively. The Daya Bay exclusion bound in the region Δm^2_{41}=0.1 to 1 eV^2 is found to be

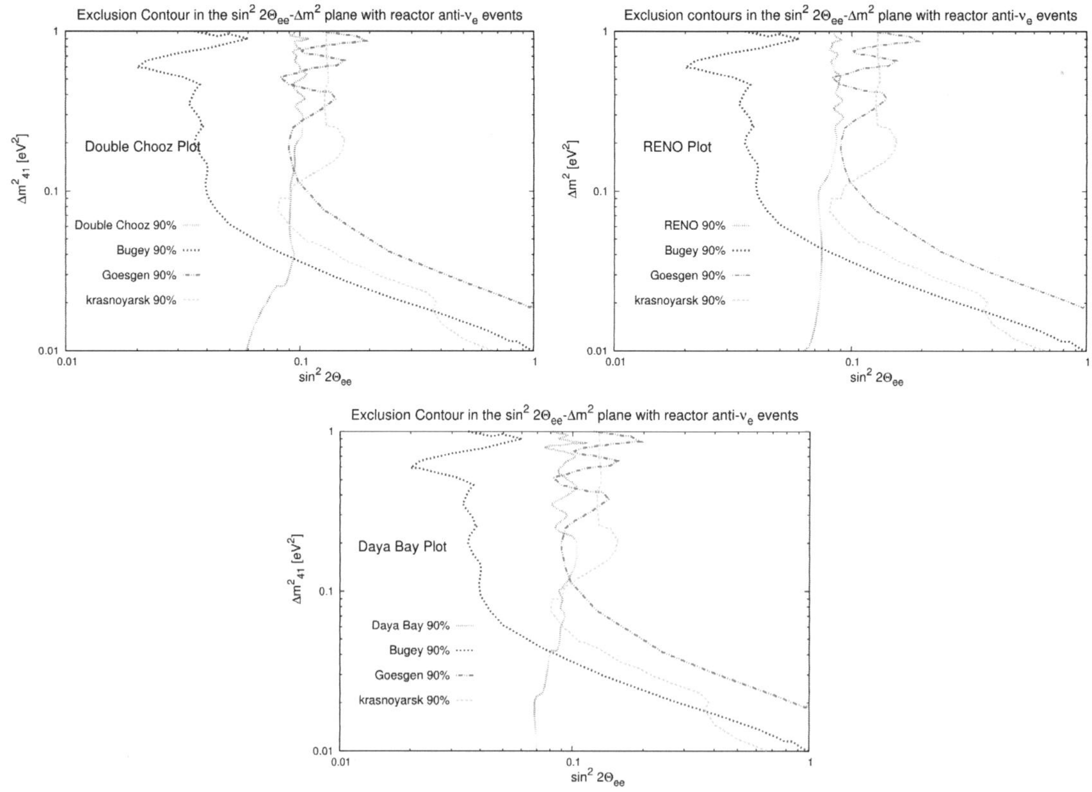

Figure 2. Comparison of Double Chooz, RENO and Daya Bay exclusion plot with other reactor experiments at 12% resolution using modified errors.

nearly $\sin^2 2\theta_{ee} > 0.07$.

5. Conclusions

From the above results, we can draw the following conclusions:

- The sensitivity of reactor experiments like Double Chooz, Daya Bay and RENO to sterile oscillation parameters is significantly dependent on the systematic errors, detector resolutions and uncertainties of each experiment. The dependence is especially strong on the correlated reactor-related errors and the normalisation uncertainty.
- Because of the multiple detectors and baselines in each of these experimental set-ups, it is possible to have partial cancellations of the experimental errors, especially of the correlated errors, which is beneficial in giving us better parameter sensitivities.
- In an analysis with duly reduced values of errors, it is possible to obtain better bounds with these set-ups than those from many of the older reactor experiments, in spite of the fact that the relatively higher baselines in this case are less suited for a determination of oscillations in the $\Delta m_{14}^2 = 1$ eV^2 range. For the latter reason, we find that we obtain better bounds in the low region $0.01 < \Delta m_{14}^2 < 0.05$ eV^2.

The appearance and disappearance experiments indicates that there is still significant uncertainty on the favoured region for sterile oscillation. In view of this, these results are significant.

6. Acknowledgements

I would like to thank Raj Gandhi and Thomas Schwetz for extensive discussions and suggestions. I thank the XI Plan Neutrino Project at HRI, Allahabad, and UGC SAP program (to Physics Department, Gauhati University) for providing financial assistance.

References

[1] A. Aguilar et al. (LSND), Phys. Rev. D64, (2001) 112007, arXiv:hep-ex/0104049.
[2] A. A. Aguilar-Arevalo et al. (MiniBooNE), Phys. Rev. Lett. 105, (2010) 181801, arXiv:hep-ex/1007.1150.
[3] G. Mention et. al., Phys. Rev. D83, (2011) 073006, arXiv:hep-ph/1101.2755.
[4] GALLEX, P. Ansetmann et. al., Phys. Lett B342, (1995) 440.
[5] SAGE, J. N. Abdurashitov et. al., Phys. Rev. C80, (2009) 015807, arXiv: hep-ph/0901.2200.
[6] J. Hamann, S. Hannestad, G. G. Raffelt, I. Tamborra, and Y. Y. Wong, Phys. Rev. Lett. 105, (2010) 181301, arXiv:hep-ph/1006.5276.
[7] E. Giusarma et al., Phys. Rev. D83, (2011) 115023 , arXiv:astro-ph/1102.4774.
[8] C. Guinti and M. Laveder, arXiv:hep-ph/1111.1069.
[9] C Guinti and M. Laveder, arXiv:hep-ph/1109.4033.
[10] R. Gandhi and P. Ghoshal, arXiv:hep-ph/1108.4360.
[11] Double CHOOZ collaboration, F. Ardellier et al., hep-ex/0606025 .
[12] A. Bandhyopadhyay and S. Choubey, arXiv: hep-ph/0707.2481.
[13] Daya Bay proposal, arXiv:hep-ex/0701029.
[14] A. de Gouvea and T. Wytock, arXiv:0809.5076.
[15] B. Bhattarcharya, Arun M. Thalapillil, C. E. M. Wagner , arXiv:hep-ph/1111.4225.
[16] D. A. Dwyer et. al., arXiv:hep-ph/1109.6036.
[17] A. B. Sousa, arXiv:hep-ph/1110.3455.
[18] A. Palazzo, Phys. Rev. D 83, (2011) 113013, arXiv:hep-ph/1105.1705.
[19] J. Evslin, H. Li, E. Ciuffoli, arXiv:hep-ph/1205.5499.
[20] J. K. Ahn, S. R. Baek, S. Choi, arXiv:hep-ph/1003.1391.
[21] Bugey, B. Achkar et al., Nucl. Phys. B434, (1995) 503 .
[22] Gosgen, G. Zacek et al., Phys. Rev. D 34, (1986) 2621.
[23] Krasnoyarsk, G. S. Vidyakin et al., Sov. Phys. JETP 71, (1990) 424.
[24] M. Maltoni, T. Schwetz, M. Tortola, and J. Valle, New J. Phys. 6, (2004) 122, arXiv:hep-ph/0405172.
[25] P. Huber, J. Kopp, M. Lindner, M. Rolinec, and W. Winter, JHEP 05, (2006) 072, arXiv:hep-ph/0601266.
[26] P. Huber, J. Kopp, M. Lindner, M. Rolinec, and W. Winter, arXiv:hep-ph/0701187.
[27] P. Huber, M. Lindner, T. Schwetz and W. Winter, arXiv:hep-ph/0907.1896.

Universal extra dimensions : life with BLKTs

Anindya Datta[1], Ujjal Kumar Dey[2], Amitava Raychaudhuri[3] and Avirup Shaw[4]

[1,3,4] Department of Physics, University of Calcutta, 92, Acharyya Prafulla Chandra Road, Kolkata 700009;
[2] Harish-Chandra Research Institute, Chhatnag Roag, Jhunsi Allahabad 211019

E-mail: [1]adphys@caluniv.ac.in, [2]ujjaldey@hri.res.in, [3]palitprof@gmail.com, [4]avirup.cu@gmail.com

Abstract. In universal extra dimension (UED) models with one compactified extra dimension, a $\mathbf{Z}_2$ symmetry, termed KK-parity, ensures the stability of the lightest Kaluza-Klein particle (LKP) which could be a viable dark matter candidate. This symmetry leads to two fixed points in the extra space like direction. In non-minimal versions of UED boundary-localized kinetic terms (BLKT) of same strength at both fixed points induce a new $\mathbf{Z}_2$ symmetry which ensures the stability of LKP. The precision of the dark matter measurements severely correlates and restricts the BLKT parameters of gauge bosons and fermions. Furthermore, BLKT parameters of different strengths at the fixed would induce a non-consevation of KK-parity. We examine, in the presence of such terms, single production and decay of Kaluza-Klein excitations of the neutral electroweak gauge bosons in the context of LHC.

1. Introduction

After the discovery of Higgs boson at the LHC, much activity is now aimed to discover the physics which lies beyond the Standard Model. The evidence for such physics, though indirect, can be traced to the issue of naturalness of the Higgs mass, the observed masses and mixing of neutrinos, and the quest for a dark matter candidate. The energy scale for new physics remains unknown but there are several motivations which encourage us to expect that it may well be within the reach of the LHC. Models in which all the Standard Model fields propagate in one or more space like extra dimensions have many attractive features and they were proposed to overcome some of the shortcomings of the SM. In this article we will be discussing one such example, namely the UED model. In the next section we will briefly introduce the model and see that a nice feature of this model is the presence of a particle which could be a candidate for dark matter. Section 2, would be devoted to UED in the light of direct and indirect evidences from DM. This will be followed by a section devoted to the signatures of this model at the LHC. Finally we conclude in section 4.

2. Overview of UED with BLKT

Our concern here is a particularly interesting framework, called the UED scenario, characterized by a single flat extra dimension, compactified on an S^1/Z_2 orbifold (with radius of compactification, R)[1]. This extra space like dimension is accessed by all the SM particles. From a 4-dimensional viewpoint, every field in the SM will then have an infinite tower of Kaluza-Klein (KK) modes, each mode being identified by an integer, n, called the KK-number. The zero modes ($n = 0$) are identified as the corresponding SM states. The orbifolding is essential to ensure that fermion zero modes have a chiral representation. But it has other consequences too. The physical region along the extra direction y is now smaller $[0, \pi R]$ than the periodicity $[0, 2\pi R]$, so the KK number (n) is no longer conserved. What remains actually conserved is the even-ness and odd-ness of the KK states, ensured through the conservation of KK parity, defined by $(-1)^n$. Lorentz invariance along this direction is also lost due to compactification, and as a result the KK masses receive bulk and orbifold-induced radiative corrections [2]. As the theory under consideration lacks ultra-violet completeion, so instead of actually estimating the radiative correction (which is any case, logarithmically sensitive to the cut-off) one might consider kinetic and mass terms localized at the fixed points to parametrise these unknown corrections. We restrict ourselves to boundary-localized kinetic terms only [3, 4, 5, 6, 7, 8, 9]. Specifically we consider a five-dimensional theory with additional kinetic terms localized at the boundaries at $y = 0$ and $y = \pi R$. We call this incarnation as non-minimal UED (nmUED).

We illustrate the idea by considering free gauge fields $G^M(x, y)$ whose zero mode may be identified with photon for example.

The action with boundary kinetic terms can be written as

$$S = -\frac{1}{4}\int d^4x \, dy \left[F_{MN}F^{MN} + r_g^a\delta(y)F_{\mu\nu}F^{\mu\nu} + r_g^b\delta(y - \pi R)F_{\mu\nu}F^{\mu\nu}\right], \quad (1)$$

where $F_{MN} = (\partial_M G_N - \partial_N G_M)$ and r_g^a, r_g^b, the strengths of the boundary terms which are free parameters in our analysis along with the compactification radius R. It is straightforward to show that in the $G_4 = 0$ gauge, the gauge field has the KK-expansion

$$G_\mu(x, y) = \sum_{n=0}^{\infty} G_\mu^n(x)a^n(y). \quad (2)$$

where the functions $a^n(y)$ are of y-dependent part of the gauge fields. In this case the five-dimensional contributions to the KK-gauge field mass, m_n satisfy

$$(r_g^a r_g^b m_n^2 - 4)\ \tan(m_n \pi R) = 2(r_g^a + r_g^b)m_n\ . \quad (3)$$

In the above, $1/R$ is the inverse of the compactification radius of the extra space like dimension and this is the only (mass) dimensionfull parameters of our analysis.

KK-masses and wave functions of fermions and scalars are obtained by a similar method and for further details we refer [9].

3. nmUED via dark matter window

One of the burning issues of the SM is absence of a viable DM candidate . Although initially neutrinos or axions were hoped to be the required DM particle, from present day cosmological observations they are disfavored. The requirement of a dark matter candidate has been one of the important motivations to go beyond SM. UED, an alternate extension of the SM, also predicts its own dark matter candidate. Unlike the minimal UED, where $n = 1$ KK-excitation of $U(1)$ gauge field , B^1 is the LKP, in nmUED, one has the choice of LKPs from B^1, H^1 or W_3^1.

However in the following analysis we consider the 1st excited KK-level of hypercharge gauge boson B^1 is the relic particle, so the all other particles are unstable and ultimately decay to B^1 in association with SM particles. For extensive informations of relic density formulation in UED the reader is referred to [11, 12].

Evolution of number density n of relic (dark matter) particle in an expanding universe is governed by the Boltzmann equation,

$$\frac{dn}{dt} = -3Hn - \langle \sigma_{eff} v \rangle (n^2 - n_{eq}^2), \tag{4}$$

here H is the Hubble parameter and n_{eq} the number density at thermal equilibrium.

$\langle \sigma_{eff} v \rangle$ ($\sim a_{eff}(x) + b_{eff}(x)\, v^2 + \mathcal{O}(v^4)$) is the thermally averaged relative velocity times the effective interaction cross section of B^1 with other particles in the spectra.

σ_{eff} can be defined as,

$$\sigma_{eff}(x) = \frac{1}{g_{eff}^2} \sum_{ij}^{N} \sigma_{ij} F_i F_j. \tag{5}$$

with

$$F_i(x) = g_i (1 + \Delta_i)^{3/2} \exp(-x\Delta_i) \text{ and } g_{eff}(x) = \sum_{i=1}^{N} F_i, \tag{6}$$

where

$$\Delta_i = \frac{m_i - m_0}{m_0} \text{ and } x = \frac{m_0}{T}. \tag{7}$$

g_i is the number of degrees of freedom for any particle which takes part in the annihilation or coannihilation process, σ_{ij} is the corresponding cross section and m_0 is the mass of relic particle.

The approximate formula for the relic density,

$$\Omega h^2 \approx \frac{1.04 \times 10^9/1\text{GeV}}{M_{Pl}} \frac{x_F}{\sqrt{g_*(x_F)}} \frac{1}{I_a + 3I_b/x_F}, \tag{8}$$

M_{Pl} is the Planck mass and $I_a = x_F \int_{x_F}^{\infty} a_{eff}(x) x^{-2} dx$, $I_b = 2x_F^2 \int_{x_F}^{\infty} b_{eff}(x) x^{-3} dx$,

here x_F is the freeze-out temperature and $g_*(x_F)$ accounts for the relativistic degrees of freedom at the freeze-out temperature.

The strengths of BLKTs have been be chosen such that B^1 is always the LKP [13]. We assume BLKTs for other particles in such a manner that coannihilation of only left-handed lepton doublet fermions are relevant. Coannihilation becomes less important as the mass splitting in the $n = 1$ level increase.

In fig.1, we present the main results of this section. For several choices of fermion BLKT parameters, Ωh^2 has been plotted as a function of $1/R$. The blue (shaded) band represents the 1σ allowed range of relic density measured from PLANCK [14].

Range BLKT parameters those are allowed from the measured relic-density can be read off from the plots by looking at the intersection of the lines with the blue band in fig.1. Unlike UED, the allowed range of R^{-1} depends on R_f in this non-minimal version of the model. As R_f moves away from R_B the splitting, Δ_f , increases and coannihilation becomes less important.

4. Probing nmUED via KK-parity violation at the LHC

In this section we will be interested in a special situation where the strength of the BLKT parameters at the two fixed points are not the same. This results into violation of KK-parity which in turn is manifested by non-vanishing couplings among $n = 0$ KK-states with $n = 1$ KK-states. In particular we will be interested in single production of B^1 and W_3^1 in proton proton collision at the LHC, followed by the decay of B^1 into a pair of SM charged leptons. Both the production and decay are driven by couplings which do not respect KK-parity.

Relevant couplings can be easily obtained in [9]. In fig.2 such KK-parity violating couplings between B^1 or W_3^1 to a pair of SM ($n = 0$ KK-mode) fermions has been plotted with ΔR (the difference between the BLKT parameters of the gauge bosons at two fixed points). As expected, these couplings grow with ΔR, the measure of KK=parity violation.

Journal of Physics: Conference Series **481** (2014) 012006 doi:10.1088/1742-6596/481/1/012006

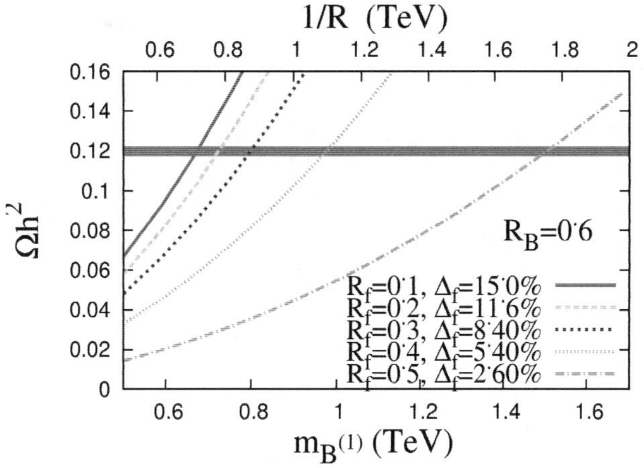

Figure 1. Variation of Ωh^2 with relic particle mass. Curves for different choices of the fermion BLKT parameter R_f are shown and the corresponding Δ_f indicated. The narrow horizontal blue band corresponds to the 1σ allowed range of relic particle density from Planck data. The allowed $1/R$ can be read off from the intersections of the curves with the allowed band. This panels correspond to a fixed value of R_B.

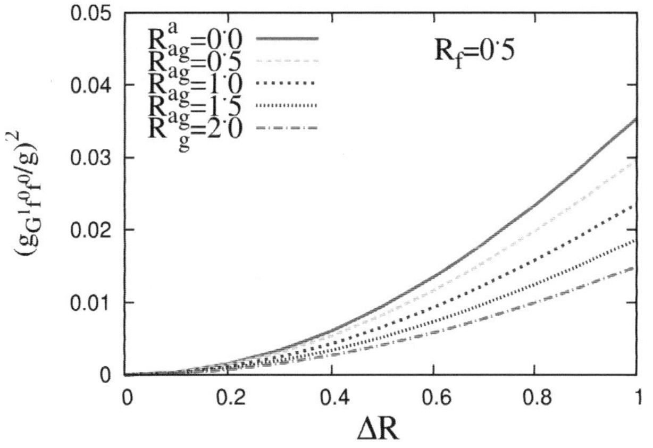

Figure 2. Variation of the square of the KK-parity violating coupling (between $B^1 or W^1_3$ and a pair of zero-mode fermions) with ΔR for several choices of R^a_g. This panels correspond to a fixed value of R_f.

We are now in a position to discuss some phenomenological signals of nmUED. In the following we will restrict our discussion only to the prospects at the LHC. Though for B^1 and W^1_3 we choose different BLKT strengths we keep ΔR to be the same for both B^1 and W^1_3. At the LHC we are interested to investigate the resonant production of the $n = 1$ KK-excitations of the neutral electroweak gauge bosons, via the process $pp(q\bar{q}) \to G^1$ followed by $G^1 \to l^+l^-$ where G^1 is either of B^1 and W^1_3 and l could be either $e^\pm$ or $\mu^\pm$. If in future such a signature is observed at the LHC, then it would be a good channel for the determination of such KK-parity violating couplings.

An analytic expression for the production cross section in proton proton collisions can be written in a compact form :

$$\sigma(pp \to G^1 + X) = \frac{4\pi^2}{3m^3_{g^{(1)}}} \sum_i \Gamma(G^1 \to q_i\bar{q}_i) \int_\tau^1 \frac{dx}{x} \left[f_{\frac{q_i}{p}}(x, m^2_{g^{(1)}}) f_{\frac{\bar{q}_i}{p}}(\tau/x, m^2_{g^{(1)}}) + q_i \leftrightarrow \bar{q}_i \right] \tag{9}$$

Here, q_i and $\bar{q}_i$ stand for a generic quark and the corresponding antiquark of the i-th flavour respectively. $\Gamma(G^1 \to q_i\bar{q}_i)$ represents the decay width of G^1 into a quark and antiquark pair of the ith flavour. $\tau \equiv m^2_{g^{(1)}}/S_{PP}$, where $\sqrt{S_{PP}}$ is the proton proton centre of momentum energy. The fs are quark or antiquark distribution functions within a proton.

In case of of B^1 production $\Gamma = \left[(g^2_{G^1 q\bar{q}}/32\pi) \times [(Y^q_L)^2 + (Y^q_R)^2]\right] m_{B^{(1)}}$ (with Y^q_L and Y^q_R being the weak hyper-charges for the left- and right-chiral quarks), while for W^1_3 one has $\Gamma = \left[g^2_{G^1 q\bar{q}}/128\pi\right] m_{W_3^{(1)}}$, $g_{G^1 q\bar{q}}$ is the KK-parity violating coupling among SM quarks. To obtain the numerical values of the cross-sections, we use a parton level Monte Carlo code with parton distribution functions as parametrized in CTEQ6L [15]. We take the pp centre of momentum energy to be 8 TeV. Renormalisation (for α_s) and factorisation scales (in the parton distributions) are set at $m_{g^{(1)}}$.

In fig.3, the main results of our analysis has been presented in the context of 8 TeV run of the LHC. As the above signal is practically background free, 40 signal event can be assumed to be a benchmark for discovery. In the figure, contours of 40 signal events have been plotted in the $\Delta R - (R^a_B - R^a_W)$ plane. If no such signal would be seen at the LHC, parameter space above a particular line could be excluded.

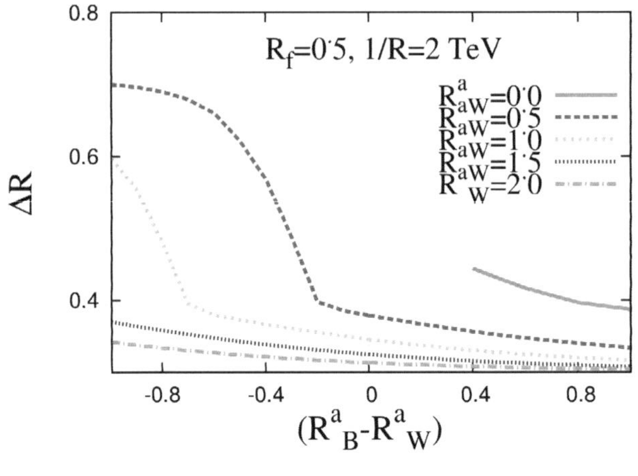

Figure 3. Iso-event curves (40 signal events with 20 fb^{-1} data at the LHC running at 8 TeV) for combined W^1_3 and B^1 signals in the $\Delta R - (R^a_B - R^a_W)$ plane for several choices of R^a_W. This panel corresponds to specific values of R_f and R^{-1} while ΔR_g is taken to be the same for W^1_3 and B^1. The regions below the curves correspond to less than 20 events for the chosen R_f and R^a_W. R^{-1} is taken as 2 TeV for this panel.

5. Summary

In summary, we have investigated phenomenology of a model with gauge group $SU(3)_c \times SU(2)_L \times U(1)_Y$, in which all the Standard Model fields propagate in $1+4$ dimensional manifold.

The extra space like dimension is compactified on an orbifold with two fixed points. Effective 4-dimensional theory thus consists of towers of excitation of SM fields. An imposed Z_2 symmetry results into presence of a weakly interacting stable massive particle in the spectra. Radiative corrections to the masses and the couplings in this theory are log sensitive to the cut-off thus having certain amount of arbitrariness. In an incarnation of this model, log divergent radiative corrections are parametrised by adding all possible kinetic terms involving the fields at the orbifold fixed poiints. We have reported the results of relic density calculation in the framework of this model. Furthermore, taking boundary kinetic terms with unequal strengths would result into some dramatic signatures of this model at colliders. In particular we have investigated production and decay of lightest KK-excitation of electroweak gauge bosons at the LHC. Observation/non-observation of such signals at the LHC in near future would definitely help us to discover/exclude nmUED.

AD thanks the organisers of NCHEPC for their warm hospitality. Research of AD is partially supported by the UGC-DRS programme at the Department of Physics, University of Calcutta. UKD is supported by funding from the DAE, Govt. of India for RECAPP, HRI. AR is partially funded by the J. C. Bose Fellowship, DST. AS thanks the UGC for support.

References

[1] Appelquist T, Cheng T and Dobrescu B, (2001), Phys. Rev. D **64** 035002.
[2] Cheng T, Matchev K and Schmaltz M, (2002), Phys. Rev. D **66** 056006.
[3] Dvali G, Gabadadze G, Kolanovic M and Nitti F, (2001), Phys. Rev. D **64** 084004.
[4] Carena M, Tait T and Wagner C, (2002) Acta Phys. Polon. B **33** 2355.
[5] del Aguila F, Perez-Victoria M and Santiago J, (2003) J. High Energy Phys. **02** 051.
[6] del Aguila F, Perez-Victoria M and Santiago J, (2003) Acta Phys. Polon. B **34** 5511.
[7] Flacke. T, Menon. T and Phalen. D, (2009) Phys. Rev. D **79** 056009.
[8] Datta A, Nishiwaki K and Niyogi S, (2012) J. High Energy Phys. **11** 154.
[9] Datta A, Dey U K, Raychaudhuri A, and Shaw A, Phys. (2013) Rev. D **87** 076002.
[10] Schwinn C, (2004) Phys. Rev. D **69** 116005 [arXiv:hep-ph/0402118].
[11] Servant G and Tait T, (2003) Nucl. Phys. B **650** 391.
[12] Kong K and Matchev K, (2006) J. High Energy Phys. **01** 038.
[13] Datta A, Dey U K, Raychaudhuri A, and Shaw A, (2013) Phys. Rev. D **88** 016011 .
[14] Planck Collaboration, Ade P et al.,e-Print: arXiv:1303.5076 [astro-ph.CO].
[15] Pumplin J *et al.*, (2002) J. High Energy Phys. **07** 012.

A discrepancy observed in the dipole anisotropy in the radio sky at 1.4 GHz and that in the CMBR – A threat to the cosmological principle?

Ashok K Singal

Astronomy and Astrophysics Division, Physical Research Laboratory, Ahmedabad 380009, India.

E-mail: asingal@prl.res.in

Abstract. According to the cosmological principle, the sky brightness at any frequency should appear uniform in all directions to an observer considered to be fixed in the co-moving coordinate system of the expanding universe. However a peculiar motion of the observer introduces a dipole anisotropy in the observed sky brightness, which should be independent of the observing frequency. We have examined the angular distribution in the radio-sky brightness, i.e., an integrated emission from discrete sources per unit solid angle, from the NVSS sample containing 1.8 million discrete radio sources at 1.4 GHz. Our results give a dipole anisotropy which is in the same direction as that of the CMBR from the COBE or WMAP, but the magnitude we find is about 4 times larger at a statistically significant (about 3σ) level. A genuine difference between the two dipoles cannot arise from the observer's motion alone, and it would imply intrinsically anisotropic universe, with anisotropy changing with the epoch. This would violate the cosmological principle where isotropy of the universe is assumed for all epochs, and on which the whole modern cosmology is based upon.

1. Introduction

An observed dipole anisotropy in the Cosmic Microwave Background Radiation (CMBR), when attributed to the peculiar motion of the solar system through the universe, yields a velocity of 369 km/s in the direction $l = 264°, b = 48°$ [1,2]. A study had found a similar anisotropy in the number counts of radio sources [3]. Here we examine the presence of such an anisotropy in the sky brightness due to radio galaxies. This will provide a confirmation on the interpretation of CMBR dipole anisotropy being due to solar motion. Also CMBR provides information about the isotropy of the universe for redshift $z \sim 700$, but the radio source population refers to a much later epoch $z \sim 1$. Thus it also provides an independent check on the cosmological principle where isotropy of the universe is assumed for all epochs.

2. The source catalogue

We have used the NVSS catalogue (NRAO VLA Sky Survey [4]) for our investigations. This survey covers whole sky north of declination $-40°$, a total of 82% of the celestial sphere, at 1.4 GHz. There are about 1.8 million sources in the catalogue with a flux-density limit $S > 3$ mJy. The catalog is available in a compact form, giving for each source right ascension (RA),

Table 1. The observed dipole asymmetry in sky brightness and the construed solar velocity

S (mJy)	N	$\Delta\mathcal{F}/\mathcal{F}$ (10^{-2})	RA (°)	Dec (°)	v (10^3 km/s)
$1000 > S \geq 100$	39625	1.6 ± 0.4	154 ± 17	-08 ± 18	1.7 ± 0.6
$1000 > S \geq 50$	90360	1.7 ± 0.4	163 ± 12	-11 ± 11	1.8 ± 0.5
$1000 > S \geq 30$	153759	1.7 ± 0.3	159 ± 11	-07 ± 10	1.8 ± 0.5
$1000 > S \geq 20$	228128	1.5 ± 0.2	158 ± 10	-06 ± 9	1.7 ± 0.4
$1000 > S \geq 15$	296811	1.5 ± 0.2	157 ± 9	-03 ± 8	1.6 ± 0.4
$100 > S \geq 50$	50735	1.7 ± 0.2	194 ± 16	-19 ± 17	1.8 ± 0.6
$100 > S \geq 30$	114134	1.4 ± 0.2	169 ± 12	-05 ± 11	1.5 ± 0.4
$100 > S \geq 20$	188503	1.2 ± 0.2	165 ± 11	-02 ± 10	1.3 ± 0.4
$100 > S \geq 15$	257186	1.2 ± 0.1	162 ± 10	$+06 \pm 10$	1.3 ± 0.3

declination (Dec) and flux density (S) at 1.4 GHz in a tabular form, and is downloadable by anonymous FTP from ftp://ftp.aoc.nrao.edu/pub/software/aips/TEXT/STARS.

3. Effects of aberration and Doppler boosting

An observer moving with a velocity v, will find sources in the forward direction brighter by a factor $\delta^{1+\alpha}$, due to Doppler boosting [5], where $\delta = 1 + (v/c)\cos\theta$ is the Doppler factor, c is the velocity of light and α (≈ 0.8) is the spectral index defined by $S \propto \nu^{-\alpha}$. Here we have used the non-relativistic formula for the Doppler factor as CMBR observations indicate that $v \ll c$. Further, the increased flux density due to Doppler boosting in the forward direction will cause a telescope of a given sensitivity limit to detect comparatively a larger number of sources. With the integral source counts of extragalactic radio source population following a power law $N(>S) \propto S^{-x}$ ($x \sim 1$) , the observed number of sources in any given flux-density range will therefore change [5] by a factor $\propto \delta^{x(1+\alpha)}$.

Also due to the aberration of light, the apparent position of a source at angle θ will be shifted in the forward direction by an angle $(v/c)\sin\theta$, as a result there will be a higher number density per steradian in the forward direction as compared to that in the backward direction. This excess [5] in number density $\propto \delta^2$. Thus as a combined effect of Doppler boosting and the aberration, the observed sky brightness (an integrated emission from discrete sources per unit solid angle) will vary as $\propto \delta^{2+x(1+\alpha)}$ which (for $v \ll c$) can be written as $1 + \mathcal{D}\cos\theta$, a dipole anisotropy over the sky [5,6] with amplitude $\mathcal{D} = [2 + x(1+\alpha)](v/c)$.

Let $\mathbf{r}_i$ be the position vector (of unit magnitude) of ith source of observed flux density S_i in the sky. Then for a uniform angular distribution of the sky brightness, $\Sigma S_i \mathbf{r}_i = 0$. However because of the observer's peculiar motion in an otherwise isotropic universe, the vectorial sum $\Sigma S_i \mathbf{r}_i$ will yield a net vector in the direction of motion, thereby fixing the direction of the dipole. If θ_i is the polar angle of the ith source with respect to the dipole direction, then the magnitude of the vectorial sum can be written as $\Delta\mathcal{F} = \Sigma S_i \cos\theta_i$. Writing $\mathcal{F} = \Sigma S_i |\cos\theta_i|$ and converting the summation into an integration over the sphere, we get for the dipole magnitude,

$$\frac{\Delta\mathcal{F}}{\mathcal{F}} = k\,\frac{\int_0^{\pi}(1+\mathcal{D}\cos\theta)\cos\theta\,\sin\theta\,\mathrm{d}\theta}{2\int_0^{\pi/2}\cos\theta\,\sin\theta\,\mathrm{d}\theta} = \frac{2k\mathcal{D}}{3} = \frac{2k}{3}\,[2 + x(1+\alpha)]\,\frac{v}{c}. \tag{1}$$

The formula is equally valid for samples with finite upper and lower flux-density limits. Here k is a constant of the order of unity ($k = 1$ for a sky fully covered by the sample) and as such may need to be determined numerically for individual cases when there are finite gaps in the sky coverage.

Table 2. The velocity vector from the number counts

S (mJy)	N	$\mathcal{D}$ (10^{-2})	RA (°)	Dec (°)	v (10^3 km/s)
≥ 50	091597	2.1 ± 0.5	171 ± 13	-18 ± 14	1.7 ± 0.4
≥ 40	115837	1.8 ± 0.4	158 ± 12	-19 ± 12	1.4 ± 0.4
≥ 35	132930	1.9 ± 0.4	157 ± 11	-12 ± 11	1.5 ± 0.3
≥ 30	154996	2.0 ± 0.4	156 ± 11	-02 ± 10	1.6 ± 0.3
≥ 25	185474	1.8 ± 0.4	158 ± 11	-02 ± 10	1.4 ± 0.3
≥ 20	229365	1.8 ± 0.3	153 ± 10	$+02 \pm 10$	1.4 ± 0.3
≥ 15	298048	1.6 ± 0.3	149 ± 09	$+15 \pm 09$	1.3 ± 0.2

As the NVSS catalogue has a gap of sources for Dec $< -40°$, in that case our assumption of $\Sigma S_i \mathbf{r}_i = 0$ for a stationary observer does not hold good. However if we drop all sources with Dec $> 40°$ as well, then with equal and opposite gaps on opposite sides $\Sigma S_i \mathbf{r}_i = 0$ is valid for a stationary observer. Further we also excluded all sources from our sample which lie in the galactic plane ($|b| < 10°$) as the excess of galactic sources towards the galactic centre is likely to contaminate the determination of velocity. Of course exclusion of such strips, which affect the forward and backward measurements identically, to a first order do not have systematic effects on the results [5]. Using Monte–Carlo simulations we estimated $k \sim 1.1$ (Eq. (1)), for the effect of the gaps in our samples.

4. Results and discussion

Our results are presented in Table.1, which is almost self-explanatory. As a relatively small number of strong sources at high flux-density levels could introduce large statistical fluctuations in the sky brightness, we have restricted our sample to below 1000 mJy level. At the lower end we have restricted it to 15 mJy levels as the completeness of the sample at weaker flux-density levels could be doubtful [3]. The estimation of errors for $\Delta\mathcal{F}/\mathcal{F}$ are described in Ref. [7].

From Table.1 the direction the velocity vector (with our best estimate RA$= 157° \pm 9°$, Dec$= -03° \pm 8°$ or in galactic co-ordinates $l = 248° \pm 12°, b = 44° \pm 8°$) is quite in agreement with those determined from the CMBR (RA$= 168°$, Dec$= -7°$ or $l = 264°, b = 48°$ with errors less than a degree) [1,2]. However the estimates of the magnitude of the velocity vector (1600 ± 400 km/s) appear much higher than the CMBR value (369 ± 1 km/s) by a factor ~ 4 at a statistically significant ($\sim 3\sigma$) level.

A previous attempt [3] had used the number counts of radio galaxies to find peculiar velocity which seemed consistent with that from the CMBR observations, but from the sky brightness anisotropy we got the velocity ~ 4 times the CMBR value. To ascertain that the difference somehow is not between the dipoles arising from the sky brightness and the number counts, we have determined the velocity from the number counts as well, using a technique [6] slightly different from that of Ref. [3]. First the direction of the dipole was determined from $\Sigma \mathbf{r}_i$, the three vector components (x, y, z) obtained essentially the same as the three $l = 1$ spherical harmonic coefficients giving the dipole [3]. Then the dipole magnitude was calculated from the fractional difference $\Delta\mathcal{N}/\mathcal{N} = \Sigma \cos\theta_i \;/\; \Sigma|\cos\theta_i| = (2k/3)[2 + x(1+\alpha)](v/c) = 2k\mathcal{D}/3$, similar to that for $\Delta\mathcal{F}/\mathcal{F}$ in the case of sky brightness. The results are summarized in Table.2. Comparing with Table.1 we notice that the observed anisotropies in both the sky brightness and the number counts yield similar velocities with magnitudes ~ 4 times the CMBR value in both cases.

To conclusively eliminate the possibility of some local clustering on our results, we determined

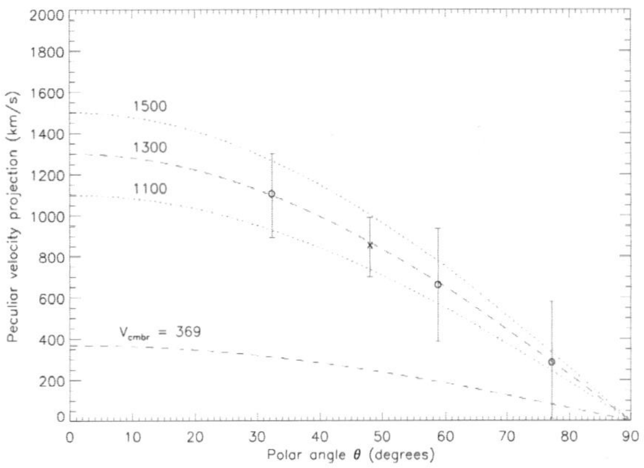

Figure 1. Observed peculiar velocity component for different polar angles. Circles (○) represent values for velocity components obtained for three different angle from six equal area slices of the sky. Cross (×) marks the valueS obtained from the whole sky, which provides the dipole magnitude for the corresponding (> 15 mJy) bin in Table.2. The CMBR value ($V_{\rm cmbr} = 369$ km/s) is shown for a comparison. The dotted lines represent the 1σ error limits about expected component values for the 1300 km/s speed.

the velocity components for three different polar angles for the last flux-density bin (> 15 mJy) which had the largest number of sources in Table.2. Starting from the dipole direction we first divided the sky into six equal-area zones. Then computing fractional difference in source counts between symmetrically placed pairs of sky zones, we determined the peculiar velocity components for three different polar angles. Fig. 1 shows a plot of the the three components, which seem to fit very well with the expected $\cos\theta$ variation of the 1300 km/s speed for that bin. Also plotted is the projection of the CMBR value (369 km/s) for a comparison. A local clustering in certain regions of the sky could not have influenced the three velocity components by the same factor and this indeed is a clinching evidence that some local clustering is not the cause of the inferred high velocity values.

Our unexpected results have recently been verified by two independent groups [8,9]. Assuming that the CMBR dipole estimates do not suffer from any residual errors during subtraction of galactic and other contributions, one cannot escape the conclusion that there is a genuine discrepancy in the two dipoles and that the reference frame defined by the radio source population at $z \sim 1$ does not coincide with that defined by the CMBR originating at $z \sim 700$. The implication are therefore serious as any differential perturbations to the Hubble flow on the scale of radio galaxies vis-á-vis that of CMBR would be truly on a universal scale. A differential anisotropy between two frames would imply anisotropic universe where anisotropy itself is changing with epoch. This would violate the cosmological principle where the isotropy of the universe is assumed for all epochs, and on which the whole modern cosmology rests upon.

References

[1] Lineweaver C H et al 1996 *Astrophy. J.* **470** 38
[2] Hinshaw G et al 2009 *Astrophy. J. Supp. Ser.* **180** 225
[3] Blake C and Wall J 2002 *Nature* **416** 150
[4] Condon J J et al 1998 *Astron. J.* **115** 1693
[5] Ellis G F R and Baldwin J E 1984 *Mon. Not. R. Astron. Soc.* **206** 377
[6] Crawford F 2009 *Astrophy. J.* **692** 887
[7] Singal A K 2011 *Astrophy. J.* **742** L23

Journal of Physics: Conference Series **481** (2014) 012007 doi:10.1088/1742-6596/481/1/012007

[8]RubartMandSchwarzDJ2013Astr.andAstrophys.555A117

[9] Kothari R, Naskar A, Tiwari P, Nadkarni-Ghosh S and Jain P 2013 *preprint* arXiv:1307.1947v2

Deflection of light due to rotating mass - a comparison among the results of different approaches

Sarani Chakraborty and Asoke kumar Sen

Department of Physics, Assam University, Silchar, Assam, India

E-mail: sarani.cool88@gmail.com

Abstract. It is known that light gets deflected due to mass. Value of deflection of light for static body is not same as it is in the case of rotating body. The deflection for a static body entirely depends on the gravitational mass where in case of a rotating body new terms will be included due to rotation. The bending angle of light is also not same in the equatorial plane and non equatorial plane for rotating Kerr body. The light bending angle is also direction of motion dependent i.e. if the motion of the light ray is in the direction of rotation, bending angle is greater than the static case and if the ray is in the opposite direction of rotation, the bending angle is smaller than the static case in equatorial plane. There are two approaches to obtain the bending angle, null geodesic of photon and change of effective refractive index. In this paper a comparison will be made among the results of different approaches.

1. Introduction

General relativity, a theory of gravitation was developed by Einstein from 1907 and finally formulated it in 1915. Two very important consequences of general relativity are bending of light ray in in presence of gravitational field and rotating mass drag the space time around them, a phenomenon called frame dragging. The exact solution of Einstein's field equation for a static, uncharged body was found by Schwarzschild in 1915, for an uncharged rotating body was solved by Kerr in 1963 and for a rotating charged body was found by Newman which is known as Kerr-Newman metric [1, 2].

So there are three factors namely mass, angular momentum and charge which can influence the curvature of space time. As a direct consequence of it, the amount of deflection of light ray would also be influenced by these factors. Many works have been done on the deflection of light due to static, rotating and charged rotating body by using two approaches, null geodesic of photon approach and effective refractive index of material medium approach. In this paper a comparison will be made among the results of different approaches to obtain the bending angle for static and rotating mass.

2. Bending of light due to static mass using null geodesic approach

In the year of 1911, a prediction was published by Einstein for bending angle of light ray from a distance source passing close to the sun at minimum distance R would get deflected with an angle.

$$\alpha = \frac{4GM}{c^2R}, \tag{1}$$

where G is gravitational constant, M is the mass of sun and c is the speed of light.

The observational verification of his prediction was made in the year 1919 during total solar eclipse. For the sun the angle of deflection of light was 1.75 sec. [1, 2]. The light deflection due to a static charged body was obtained by Virbhadra, Narasimher & Chitre up to second order term [4]

$$\alpha = \frac{4GM}{c^2 r_0} + \frac{4G^2M^2}{C^4 r_0^2}\left(\frac{15\pi}{16} - 2\right) + \frac{2}{r_0^2}\left\{2\frac{GM}{c^2}\left[\left(\frac{GM}{c^2}\right)^2 + q^2\right]^{\frac{1}{2}} - \frac{q^2\pi}{8}\right\}, \tag{2}$$

where q is the charge of the body, the bending angle for static body would be obtained by setting q equals zero

$$\alpha = \frac{4GM}{c^2 r_0} + \frac{4G^2M^2}{C^4 r_0^2}\left(\frac{15\pi}{16} - 2\right) + \frac{4G^2M^2}{C^4 r_0^2} \tag{3}$$

Higher order terms obtained by Keeton & Petter [5]

$$\alpha = 4h + \left(-4 + \frac{15\pi}{4}\right)h^2 + \left(\frac{122}{3} - \frac{15\pi}{2}\right)h^3 + \left(-130 + \frac{3465\pi}{64}\right)h^4 + \left(\frac{7783}{10} - \frac{3465\pi}{16}\right)h^5 + \left(-\frac{21397}{6} + \frac{310695\pi}{256}\right)h^6 + \mathrm{O}(h)^7 \tag{4}$$

Where,

$$h = \frac{m_0}{r_0}$$

r_0 =closest approach and

$$m_0 = \frac{GM}{c^2}$$

In terms of impact parameter b,

$$\alpha(b) = 4\left(\frac{m_0}{b}\right) + \frac{15\pi}{4}\left(\frac{m_0}{b}\right)^2 + \frac{128}{3}\left(\frac{m_0}{b}\right)^3 + \frac{3465\pi}{64}\left(\frac{m_0}{b}\right)^4 + \frac{3584}{5}\left(\frac{m_0}{b}\right)^5 + \frac{255255\pi}{256}\left(\frac{m_0}{b}\right)^6 + \mathrm{O}\left(\frac{m_0}{b}\right)^7 \tag{5}$$

An analytical perturbation frame work was developed by Iyer & Petter [6] to obtain the bending angle of light due static black hole which is

$$\alpha = 4h + \left(-4 + \frac{15\pi}{4}\right)h^2 + \left(\frac{122}{3} - \frac{15\pi}{2}\right)h^3 + \left(-130 + \frac{3465\pi}{64}\right)h^4 + \left(\frac{7783}{10} - \frac{3465\pi}{16}\right)h^5 + \left(-\frac{21397}{6} + \frac{310695\pi}{256}\right)h^6 + \mathrm{O}(h)^7 \tag{6}$$

Equation (6) is similar with equation (4)

2.1. Bending of light due to static mass using material medium approach

Fishback & Freeman obtained the bending angle of light up to first order term for static body using effective refractive index of material medium approach [3]

$$(\varphi)^1 = \frac{2(1+\gamma)GM}{bc^2} \tag{7}$$

For $\gamma=1$ this expression is similar to Einsteins expression of bending angle. More exact expression for gravitational deflection of light had been given by Sen (2010) [9] using the material medium approach and the deflection angle for static mass was obtained as

$$\Delta\phi = 2\frac{r_0^2}{r_g(r_0-r_g)}\int_0^{\frac{r_g}{r_0}} x\left\{1-\left[\frac{r_0^2}{r_g(r_0-r_g)}\right]^2 x^2(1-x)^2\right\}^{-\frac{1}{2}} dx \tag{8}$$

Where $x = r_g / r$ and r_0=closest approach and for weak field this expression reduce to standard form

$$\Delta\varphi = \frac{4GM}{c^2 r_0} \tag{9}$$

It can be said that the results are similar for both null geodesic of photon and effective refractive index of material medium approach.

3. Bending angle of light due to rotating mass.

In general relativity the space time around a body is described by Kerr metric. According to this metric a rotating body drag the space-time around it and every object coming close to the rotating body entrained to participate in its rotation. This effect is called frame dragging. The rotational frame dragging effect was first derived from the theory of general relativity by two Austrian physicists Josef Lense and Hans Thirring [1,2], which was known as Lense-Thirring effect. The deflection produced in presence of a rotating black hole explicitly depends on direction of light. Compared to the zero spin Schwarzschild case, the bending angle was greater for direct orbit and smaller for retrograde orbits. Exact bending angle for rotating black hole in the equatorial plane obtained by Iyer & Hansen is [7, 8].

$$\alpha(b) = 4\left(\frac{m_0}{b}\right)+\left(\frac{15\pi}{4}-4sa\right)\left(\frac{m_0}{b}\right)^2+\left(\frac{128}{3}-10\pi sa+4a^2\right)\left(\frac{m_0}{b}\right)^3+ \left(\frac{3465\pi}{64}-192sa+\frac{285\pi}{16}a^2-4sa^3\right)\left(\frac{m_0}{b}\right)^4 \tag{10}$$

where s is +1 for direct and -1 for retrograde motion and

$$a = \frac{j}{Mc}$$

and j is the angular momentum.

Azami, Keeton & Petter [10, 11] have shown that for off equatorial light ray bending angle has two components. One is in the equatorial plane and another is perpendicular to the equatorial plane. Equatorial component is

$$\alpha(b) = 4\left(\frac{m_0}{b}\right) + \left(\frac{15\pi}{4} - 4sa\right)\left(\frac{m_0}{b}\right)^2 + \left(\frac{128}{3} - 10\pi sa + 4a^2\right)\left(\frac{m_0}{b}\right)^3 + \left(\frac{3465\pi}{64} - 192sa + \frac{285\pi}{16}a^2 - 4sa^3\right)\left(\frac{m_0}{b}\right)^4 \quad (11)$$

This expression in equation (11) is similar to the expression obtained by Iyer and Hansen [7] shown in equation (10). Both expression reduce to Schwarzschild series if a is set to zero.

4. Conclusions

From above discussion three points are clear that whatever is the method (null geodesic of photon or effective refractive index of material medium) the expression of light deflection is same in case of static gravitational mass. Equatorial component of quasi-equatorial bending angle is similar to equatorial bending angle and for zero spin it reduces to the bending angle expression for static body. Spin does not have any contribution to the first order term.

References

[1] Landau L D and Lifshitz E M 1980 *The Classical Theory of Fields volume 2* 4th edition Butterworth-Heinemann
[2] Weinberg S 1972 *Principles and Applications of the General Theory of Relativity* John Wiely & Sons Inc.
[3] Fishback E and Freeman B S 1980 *Phys. Rev.* D **22** 12
[4] Virbhadra K S, Narasimha D and Chitre S M 1998 *Astron. Astrophys.* **337** 1-8
[5] Keeton C R and Petters A O 2005 *Phys. Rev.* D **72** 104006
[6] Iyer S V and Petters A O 2007 *Gen. Relativ. Gravit.* **39** 1563-1582
[7] Iyer S V and Hansen E C 2009 *Phys. Rev.* D **80** 124023
[8] Iyer S V and Hansen E C 2009 *arXiv:gr-qc/0908.0085*
[9] Sen A K 2010 *Astrophysics* **53** 4
[10] Aazami A B, Keeton C R and Petters A O 2011 *J. Math. Phys.* **52** 092502
[11] Aazami A B, Keeton C R and Petters A O 2011 *J. Math. Phys.* **52** 102501

$f(R)$ gravity cosmology in scalar degree of freedom

Umananda Dev Goswami and Kabita Deka

Department of Physics, Dibrugarh University, Dibrugarh 786004, Assam, India.

E-mail: umananda2@gmail.com

Abstract. The models of $f(R)$ gravity belong to an important class of modified gravity models where the late time cosmic accelerated expansion is considered as the manifestation of the large scale modification of the force of gravity. $f(R)$ gravity models can be expressed in terms of a scalar degree of freedom by explicit redefinition of model's variable. Here we report about the study of the features of cosmological parameters and hence the cosmological evolution using the scalar degree of freedom of the $f(R) = \xi R^n$ gravity model in the Friedmann-Lemaître-Robertson-Walker (FLRW) background.

1. Introduction

Although the General Relativity (GR) is a very well-tested and widely accepted theory of gravity, the explanation of the late time accelerated expansion of Universe [1] is out of scope within it's ambit. However, because of all the successful test performed so far on GR for the low scales, attempts have been made to modify GR in order to account the late time behaviour of the cosmic expansion as a consequence of the large scale modification of gravity [2, 3]. The simplest class of modified gravity theories is the $f(R)$ gravity theories in which Einstein gravity is modified by replacing the Ricci curvature scalar R by an arbitrary curvature function $f(R)$. In the special case when $f(R) \to R$, $f(R)$ gravity converge to GR without a cosmological constant [2, 3]. In last few years the $f(R)$ gravity models have been studied extensively in the cosmological and astrophysical aspects because of a number of very interesting results [4].

It is interesting that a new scalar degree of freedom (sometimes referred as *scalaron*) appears in $f(R)$ theories of gravity due to redefinition of model's variable. The conformal transformation of the metric to the Einstein frame makes it explicit in the action [2, 4]. In $f(R)$ gravity the scalar field is coupled to non-relativistic matter (dark matter, baryons) with a universal coupling constant $-1/\sqrt{6}$, and hence it is convenient work on $f(R)$ gravity in the Einstein frame, in which a canonical scalar field is coupled to non-relativistic matter, as far as the scalar field is concerned [2, 4]. In this work we have studied the $f(R) = \xi R^n$ gravity model using its scalar degree of freedom in the Einstein frame in the Friedmann-Lemaître-Robertson-Walker background. The basic aim of this study is to see numerically the features of the scalar field coupled to the $f(R)$ gravity model and hence to see the consequent evolution of the cosmological parameters in the Einstein frame. We have considered this model because it has exact power-law solutions and as a result the singularity is not appear in it as suffered by other $f(R)$ models. This manifests in the scalar degree of freedom as the scalar field potential of this model is well behaved function of the field [5]. In Sec.2 we discuss very briefly the basics of formalism of the $f(R)$ gravity theory in the Jordan frame and then we transform the formalism to the Einstein frame using the conformal transformations to developed field equations in terms scalar degree of freedom. The numerical analysis of the work based on the formalism developed in the Sec.2 is done in the Sec.3 for our model. Finally in the Sec.4 we concludes the results of our analysis.

2. $f(R)$ gravity formalism in Einstein frame

The $f(R)$ gravity action in the Jordan frame is given by [2, 3]

$$S = \frac{1}{2\kappa^2}\int d^4x\sqrt{-g}f(R) + \int d^4\mathcal{L}_m(g_{\mu\nu}, \Psi_m), \tag{1}$$

where $\kappa^2 = 8\pi G$, g is the determinant of the metric $g_{\mu\nu}$, and $\mathcal{L}_m$ is the matter Lagrangian, which is the function of the metric $g_{\mu\nu}$ and the matter fields Ψ_m. The variation of the action (1) with respect to $g_{\mu\nu}$ leads to the equation of motion [2]:

$$f'R_{\mu\nu} - \nabla_{\mu\nu}f' + \left(\Box f' - \frac{1}{2}f\right)g_{\mu\nu} = \kappa^2 T_{\mu\nu}, \tag{2}$$

where ($\prime$) denotes the derivatives with respect to R. The term $\Box f'$ does not vanish in modified gravity, which means that there is a propagating scalar degree of freedom, defined as, $\varphi = f'$. In practice, it is usually difficult to invert the definition of the scalar degree of freedom explicitly for a given $f(R)$ and only convenient to determine effective potential in a parametric form. However, using the conformal transformation to the Einstein frame it is possible to express the scalar degree of freedom of a given $f(R)$ in the explicit form. In view this we now switch to the Einstein frame under the following conformal transformation [2]:

$$\tilde{g}_{\mu\nu} = \Omega^2 g_{\mu\nu}, \tag{3}$$

where Ω^2 is the conformal factor and the quantities in the Einstein frame are represented by a tilde over them. If we make a choice, $\Omega^2 = f'$, for $f' > 0$ and introduce a new scalar field ϕ defined by $\kappa\phi \equiv \sqrt{3/2}\ln f'$, then under the transformation (3) the action in the Einstein frame can be expressed as [2]

$$S_E = \int d^4x\sqrt{-\tilde{g}}\left[\frac{1}{2\kappa^2}\tilde{R} - \frac{1}{2}\tilde{g}^{\mu\nu}\partial_\mu\phi\partial_\nu\phi - V(\phi)\right] + \int d^4x\mathcal{L}_m(F^{-1}(\phi)\tilde{g}_{\mu\nu}, \Psi_m), \tag{4}$$

where $V(\phi) = U/f'^2 = (f'R - f)/(2\kappa^2 f'^2)$ is the potential of the field ϕ. Here the conformal factor is $\Omega^2 = f' = \exp(\sqrt{2/3}\kappa\phi)$, which is field dependent. Taking the variation of the action (4) with respect to the field ϕ, we obtain the field equations without the presence of matter as [2]

$$\ddot{\phi} + 3\tilde{H}\dot{\phi} + V_{,\phi} = 0, \quad \tilde{H}^2 = \frac{\kappa^2}{3}\left[\frac{1}{2}\dot{\phi}^2 + V(\phi)\right]. \tag{5}$$

Here the dots over ϕ indicate the derivatives with respect to conformal time (i.e. cosmic time in Einstein frame) $\tilde{t}$. $\tilde{H} = \dot{\tilde{a}}/\tilde{a}$ is the Hubble parameter in the Einstein frame, $\tilde{a}(\tilde{t})$ being the scale factor in the same frame in spatially flat Friedmann-Lemaître-Robertson-Walker (FLRW) spacetime with a comic time-dependent scale factor $a(t)$ in the Jordan frame. Relations of the cosmic time and scale factor in two frames are [2]: $d\tilde{t} = \sqrt{f'}dt$, $\tilde{a} = \sqrt{f'}a$, for $f' > 0$. We finally define the energy density and the pressure of the scalar field as [2]

$$\tilde{\rho}_\phi = \frac{\dot{\phi}^2}{2} + V(\phi), \text{ and } \tilde{P}_\phi = \frac{\dot{\phi}^2}{2} - V(\phi). \tag{6}$$

It is evident from the above formalism that in Einstein frame the scalar degree of freedom for a given $f(R)$ model can be studied explicitly and conveniently.

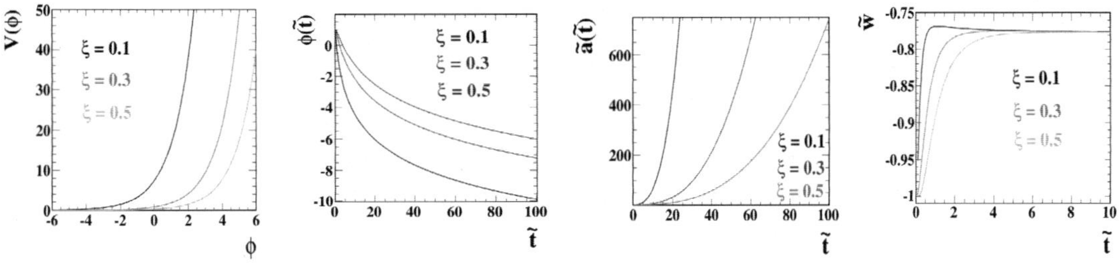

Figure 1. Few numerical results of the work. Here value of κ is taken as one. See [4] for the comprehensive work related to it.

3. Scalar field dynamics of the $f(R) = \xi R^n$ model

Using the formalism of the previous section, in this section we will study numerically different features of the scalar field and consequent cosmological implications of the model, viz.,

$$f(R) = \xi R^n, \tag{7}$$

where ξ and n are model parameters. Here our main emphasis is to see the explicit behavior of scalar degree of freedom possessed by $f(R)$ gravity model and its usefulness to study the cosmological evolution. The scalar field ϕ corresponding to the model (7) can be expressed as

$$\phi = \sqrt{\frac{3}{2}}\frac{1}{\kappa}\ln f' = \sqrt{\frac{3}{2}}\frac{1}{\kappa}\ln(\xi n R^{n-1}). \tag{8}$$

Using this expression for ϕ, the field potential takes the form

$$V(\phi) = \frac{n-1}{2\kappa^2 \xi n^2}\left(\frac{\exp(\sqrt{2/3}\kappa\phi)}{\xi n}\right)^{\frac{2-n}{n-1}}. \tag{9}$$

The first panel of the Fig.1 shows the variation of field potential with respect to the field ϕ for $n = 1.5$ with $\xi = 0.1$, 0.3 and 0.5. From this panel it is seen that the potential is well behaved function of the field and the potential is small for negative value of the field and it increases very fast as soon as the field ϕ become positive. This process is slowing down when the value of ξ increases, which is more effective for lower value range of ξ. We found that there is no significant difference of the potentials for the different values of n within the range of negative field, but the difference becomes more prominent as soon as the field ϕ acquired the positive values. The effect of n on the value of the potential is similar to that for ξ, but the effect on higher positive value of ϕ is more noticeable in this case than the case for ξ [4].

Using the potential (9), the coupled field equations (5) are solved numerically for different values of n and ξ. The results of the numerical solution are shown in the second panel of the Fig.1 for $n = 1.5$ as an example. As clear from this panel that the value of the field ϕ falls towards to the negative values starting from its initial value as time increases. For the initial period of cosmic time ϕ falls very fast and this tendency slows down as time passage. For higher values of ξ the overall tendency of the ϕ to fall towards the negative values is less in comparison to its lower values. This is similar to the nature of the field potential with respect to ξ. It is found that the field falls faster initially for the smaller value of n as compared to the higher value, but the situation reversed in the latter cosmic times. During the time before the situation is reversed the difference of the field ϕ is not very high, however after this period the difference become increasingly significant as time elapsed [4]. From these numerical solutions we have calculated the scale factors $\tilde{a}(\tilde{t})$ in Einstein frame for different values of n and ξ. The (cosmic) time evolution of the scale factors obtained from these calculation are shown in the third panel of the Fig.1 for $n = 1.5$ only. It is observed that the $f(R) = \xi R^n$ gravity model leads the very fast expansion of the Universe in the Einstein frame similar to the exponential expansion. The rate of expansion of the Universe slowed down gradually for higher values of ξ in comparison to its lower values. During initial period this process is slow, but become more significant as time passage. On the other hand the expansion is slower for lower value of n in comparison to the case for the higher value of n. In the initial period the difference of the values of the scale factor is very lass, but becomes more prominent in the latter times [4].

Finally we have calculated the equation of state $\tilde{w} = \tilde{\rho}_\phi/\tilde{P}_\phi$ in Einstein frame using equations (6) from the numerical solutions mentioned above, some of results are shown in the last panel of the Fig.1. It is seen that the equation of state $\tilde{w}$ of the scalar field ϕ for the model (7) always less than -0.75 for all cases of our study and for all times. For higher values of ξ, $\tilde{w}$ is smaller than that for smaller value of ξ during period for some initial times, but for latter times $\tilde{w}$ becomes equal for all ξ. For higher value of n, $\tilde{w}$ is always smaller than that for the smaller value of n with a similar trend with respect to time and ξ as mentioned above [4]. As the equation of state is always negative, therefore the model (7) could produced the accelerated expansion of the Universe and it corresponds to dark energy, whose equation of state is $w < -1/3$. If we increase the values of n then the equation of state $\tilde{w}$ in Einstein frame will approach to -1. This implies that the model (7) with higher values of n (< 2) corresponds to the cosmological constant Λ [4].

4. Conclusion

We have investigated the scalar degree of freedom and it's related cosmological implications of the $f(R)$ gravity model ξR^n with the different values of the model parameters ξ and n. As the scalar degree of freedom of the $f(R)$ gravity model can be treated explicitly in the Einstein frame, so we used this frame for this study by conformal transformations from the Jordan frame. It is found that the field potential is the well behaved function of the field ϕ representing the scalar degree of freedom of the model and increases very fast as ϕ increases, which depends on the model parameters ξ and n such that for the lower value of them such process is faster than higher value. From the numerical simulation of the field equations, we found that ϕ falls towards the negative values as time increases, which is more faster during the initial period of time than late times. The rate of falling of the value of the field is slower for the higher values of the parameter ξ. The study of the time evolution of the scale factor in the Einstein frame shows that the Universe is accelerating very fast for our $f(R)$ gravity model, which depends on the values ξ and n in such a way that accelerating process slows down for higher values of ξ and for lower values of n. The equation of state of the scalar field for the model is always negative and less than $-1/3$, which corresponds to the behaviour of the dark energy that produces the accelerated expansion of the Universe. The equation of state of the field for our model tends to -1 as the value of n increases towards higher values, nearly equal to 2. Thus for such values of n, the scalar field for the model behave as the cosmological constant Λ [4].

References

[1] A. G. Riess et al., Astron. J. **116**, 1009 (1998); S. Perlmutter et al., Astrophys. J. **517**, 565 (1999).
[2] A. De Felice and S. Tsujikawa, Living Rev. Rel. **13**, 3 (2010).
[3] T. P. Sotiriou and V. Faraoni, Rev. Mod. Phys. **82**, 451 (2010).
[4] U. D. Goswami and K. Deka, IJMPD **22(13)**, 1350083 (2013).
[5] N. Goheer, J. Larena, and P. K. S. Dunsby, Phys. Rev. D **80**, 061301(R) (2009).

Gravitational redshift in Kerr field

Anuj Kumar Dubey and Asoke Kumar Sen

Department of Physics, Assam University, Silchar-788011, Assam, India.

E-mail: danuj67@gmail.com

Abstract. Generally gravitational redshift has been calculated without consideration of rotation of a body. Neglecting the rotation, the geometry of space time can be described by using the well-known spherically symmetric Schwarzschild geometry. Rotation has great effect on general relativity, which gives new challenges on gravitational redshift. When rotation is taken into consideration spherical symmetry lost and off diagonal terms appears in the metric and the geometry of space time can be described by using the Kerr solution, which is the exact solution of the Einstein's field equations known at present. In this paper we will derive the expression for gravitational redshift for rotating source in Kerr field, and also apply the derived expression to calculate the gravitational redshift in case of Sun under Newtonian approximation of angular momentum.

1. Introduction

A line spectrum emitted by some atoms located, for example, on the sun, looks the same there as the spectrum emitted by the same atoms located on the earth would appear on it. If, however, we observe on the earth the spectrum emitted by the atoms located on the sun, then, its lines appear to be shifted with respect to the lines of the same spectrum emitted on the earth. Each line with frequency ω will be shifted through the interval $\Delta\omega < 0$, i.e. the shift occurs in the direction of lower frequency. The phenomenon we have described is called the "redshift". The first claimed confirmation of the predicted gravitational redshift came from the measurement of the apparent radial velocity of Sirius B, the white dwarf companion to Sirius (Adams 1925 [1]), identified as the difference from the motion expected in its 50 year orbit. The first Experimental verification of Gravitational redshift is gravitational redshift in nuclear resonance (Pound and Rebka 1959 [2]). An improved version of the experiment of Pound and Rebka has been performed to measure the effect of gravity (Pound and Snider 1965 [3]) making use of the Mossbauer Effect; the result found was (0.9990 ± 0.0076) times the value 4.905×10^{-15} of $2gh/c^2$ predicted from principle of equivalence. The redshift of the solar potassium absorption line at 7699 angstrom has been measured (Snider 1972 [4]) by means of an atomic - beam resonance - scattering technique. Gravitational redshift of sun has been also measured by (Krisher et al. 1993 [5]). A expression for gravitational redshift factor has been also calculated (P D Nunez and M Nowakowski, 2010 [6]) by applying small perturbation to the Schwarzschild geometry and two main highlighted results are the derivation of a maximum angular velocity depending only on the mass of the object and a possible estimates of the radius.

Covariant form of metric tensor in Boyer – Lindquist coordinates with signature (+,-,-,-) is expressed as

$$ds^2 = g_{tt}c^2dt^2 + g_{rr}dr^2 + g_{\theta\theta}d\theta^2 + g_{\phi\phi}d\phi^2 + 2g_{t\phi}cdtd\phi \tag{1}$$

Non-zero components $g_{\mu\nu}$ of Kerr field are given as follows

$$g_{tt} = \frac{\Delta - a^2\sin^2\theta}{\rho^2} \tag{2}$$

$$g_{rr} = -\frac{\rho^2}{\Delta} \tag{3}$$

$$g_{\theta\theta} = -\rho^2 \tag{4}$$

$$g_{\phi\phi} = -\frac{\left[(r^2+a^2)^2 - a^2\Delta\sin^2\theta\right]\sin^2\theta}{\rho^2} \tag{5}$$

$$g_{t\phi} = \frac{2mra\sin^2\theta}{\rho^2} \tag{6}$$

With

$$\rho^2 = r^2 + a^2\cos^2\theta \tag{7}$$

$$\Delta = r^2 - 2mr + a^2 = r^2 - r_g r + a^2 \tag{8}$$

Where the parameters, $m = \frac{GM}{c^2}$, e and $a = \frac{J}{Mc}$ are respectively associated with mass, charge and angular momentum per unit mass (rotation parameter) of the source. Where Schwarzschild radius, $r_g = 2m$.

For equatorial plane, $\theta = \frac{\pi}{2}$ the metric elements becomes,

$$g_{tt} = 1 - \frac{2m}{r} = 1 - \frac{r_g}{r} \tag{9}$$

$$g_{\phi\phi} = -(r^2 + a^2 + \frac{2ma^2}{r}) = -(r^2 + a^2 + \frac{r_g a^2}{r}) \tag{10}$$

$$g_{t\phi} = \frac{2ma}{r} = \frac{r_g a}{r} \tag{11}$$

2. Gravitational redshift from rotating body

The lagrangian of the test particle can be expressed as,

$$L = \frac{g_{ij}}{2}\dot{x}^i\dot{x}^j \tag{12}$$

(Dot over a symbol denotes ordinary differentiation with respect to an affine parameter λ)
From lagrangian of the test particle, we can obtain the momentum of the test particle as,

$$P_i = \frac{\partial L}{\partial \dot{x}^i} = g_{ii}\dot{x}^j \tag{13}$$

Thus corresponding momentum in the coordinates of t, r, θ and ϕ are expressed as

$$P_t \equiv -E = g_{tt}\dot{t} + g_{\phi\phi}\dot{\phi} \tag{14}$$

$$P_r = g_{rr}\dot{r} \tag{15}$$
$$P_\theta = g_{\theta\theta}\dot{\theta} \tag{16}$$
$$P_\phi \equiv L = g_{t\phi}\dot{t} + g_{\phi\phi}\dot{\phi} \tag{17}$$

Since the space time of the Kerr family is stationary and axially symmetric, and g_{ij} is a function of r and θ only, the momenta P_t and P_ϕ are conserved along the geodesics. So we obtain two constants of motion: one is corresponding to the conservation of energy (E) and the other is the angular momentum (L) about the symmetry axis.

Considering Stationary and axis symmetric space time, so a faraway observer can measure rotating body angular velocity (Ω) given by (Straumann N 1984 [7]),

$$\Omega = \frac{d\phi}{dt} = \left(\frac{d\phi}{d\tau}\right)\left(\frac{d\tau}{dt}\right) = \frac{u^\phi}{u^t} \tag{18}$$

Four velocity of a stationary point on the surface can be written as,

$$u^\mu = \left(u^t, 0, 0, u^\phi\right) \tag{19}$$

Using equation (18) in above equation (19)

$$u^\mu = \left(u^t, 0, 0, \Omega u^t\right) \tag{20}$$

Through the normalization condition of the four velocity given by (Landau LD 1980 [8])

$$u_\mu u^\mu = 1 = g_{\mu\nu} u^\mu u^\nu \tag{21}$$
$$g_{tt}u^t u^t + 2g_{t\phi}u^t u^\phi + g_{\phi\phi}u^\phi u^\phi = 1 \tag{22}$$

Using equation (20) in above equation (22) , we obtain the time-like component of the four velocities in terms of the metric components of Kerr field and angular velocity of rotation (Ω).

$$u^t = (g_{tt} + 2g_{t\phi}\Omega + g_{\phi\phi}\Omega^2)^{1/2} \tag{23}$$

Any observer measures the frequency (ω) of a photon following null geodesic $x^\mu(\lambda)$ can be calculated by the expression given by (Carroll SM 2004 [9]),

$$\omega = u^\mu \frac{dx^\mu}{d\lambda} = u^\mu g_{\mu\nu} \frac{dx^\nu}{d\lambda} \tag{24}$$

If photon is emitted at $r = \theta =$ constant, $dr = d\theta = 0$, then the frequency (ω) can be expressed as

$$\omega = u^t(g_{tt}\dot{t} + g_{t\phi}\dot{\phi}) + u^\phi(g_{t\phi}\dot{t} + g_{\phi\phi}\dot{\phi}) \tag{25}$$

Using equation (14 and 17) in above equation (25), we can write

$$\omega = u^t(-E) + u^\phi(L) = u^t(-E + \Omega L) \tag{26}$$

Using equation (23) in above equation (26), we can write the expression of frequency observed as

$$\omega = \frac{(-E + \Omega L)}{(g_{tt} + 2g_{t\phi}\Omega + g_{\phi\phi}\Omega^2)^{1/2}} \tag{27}$$

In general relativity gravitational redshift (Z) and redshift factor (g) is defined as

$$\frac{1}{Z+1} = g = \frac{\omega_{observed}}{\omega_{emitted}} \tag{28}$$

From equation (27) and equation (28), we can write the expression of gravitational redshift (Z) and gravitational factor (g) as,

$$\frac{1}{Z+1} = g = (g_{tt} + 2g_{t\phi}\Omega + g_{\phi\phi}\Omega^2)^{1/2} \tag{29}$$

We can conclude from equation (29) that the gravitational redshift will be affected by rotation of source because the expression contains the angular velocity of rotation of source in Kerr field. If we consider angular velocity (Ω) of source is zero then we can obtain the corresponding gravitational redshift from a static body of same mass.

3. Applications

Using equation (29), with the values of metric coefficients of Kerr field, we can calculate the numerical values of gravitational redshift from a rotating body.

Let us consider Sun as a rotating body, in case of Sun using the following values of Mass $(M) = 2.00\times10^{30}$, Radius $(r) = 0.70\times10^{6}$ km, Schwarzschild radius (r_g)= 3 km, Period (P)= 28 days.

Angular velocity of rotation,

$$\Omega = \frac{2\pi}{P} = 2.5982\times10^{-6}\,\text{rad/s} \tag{30}$$

Considering the Newtonian approximation of angular momentum,

$$J = \frac{2Mr^2\Omega}{5} \tag{31}$$

Rotation parameter,

$$a = \frac{J}{Mc} = \frac{2r^2\Omega}{5c} = 1.69749\text{ km} \tag{32}$$

The calculated value of gravitational redshift (Z) for Sun at equator $\theta = \frac{\pi}{2}$, is 2.11748247×10^{-6} and the corresponding value of redshift of same body (Sun) neglecting the rotation is 2.11746703×10^{-6}. The above calculation indicates that due to rotation effect there is change in gravitational redshift.

4. Acknowledgement

A Dubey is thankful to Sarani Chakraborty and Tanay Ghosh, Department of Physics, Assam University Silchar, for their useful discussions and suggestions.

References

[1] Adams W S, 1925, Observatory, **48**, 337.
[2] Pound R V and Rebka G V Jr., 1959, *Phys. Rev. Lett.,* Volume **3**, Number 9.
[3] Pound R V and Snider J L, 1965, *Physical review*, Volume **140**, Number 3B.
[4] Snider J L, *Phys. Rev. Lett.,* 1972, Volume **28**, Number 13.
[5] Krisher T P, Morabito D and Anderson J D , 1993, *Phys. Rev. Lett.,* **70**, 2213.
[6] Nunez P D and Nowakowski M, 2010, *J. Astrophys. Astr.*, **31**, 105–119.
[7] Straumann N, 1984, *General Relativity with applications to Astrophysics*, Springer Verlag.
[8] Landau L D and Lifshitz E M, 1980, *The Classical Theory of Fields,* Volume **2**, 4th edition Butterworth-Heinemann.
[9] Carroll SM, 2004, *Space time and Geometry,* Pearson Addison Wesley.

Impact of interacting dark sector on Cosmic Microwave Background Radiation

M M Verma

Department of Physics, University of Lucknow, Lucknow 226 007, India.

E-mail: sunilmmv@yahoo.com

Abstract. It has been shown earlier that an arc-like pattern found on the Cosmic Microwave Background Radiation (CMBR) may result from the decay of dark matter particles initiating near particle horizon in the Q-phase of the interacting cosmological constant (ICC) model. In the present work, an investigation is made into how the corresponding decay of such dark matter particles might influence these signatures, in view of the recent data on CMBR and the diffuse glow of the anomalous microwave radiation. We also discuss the constraints on such decay imposed by the interaction of the cosmological constant with the background. These predictions made in the ICC model can be verified in the concordance space of multiple observations.

1. Introduction

Dark energy has emerged as a major component of the universe following the observations of type Ia supernovae (SNe Ia) [1, 2] and the acoustic peak at the Legendre multipole $l_{\text{peak}} \sim 197$ in the Cosmic Microwave Background Radiation (CMBR) power spectrum [3].

We have shown previously [4] that in a three-phase universe λ must vary due to interaction (then we call it a parameter) in the intermediate phase II known as Q phase. In phase I and III, however, it remains a constant and does not interact, where it causes the required acceleration. Thus in Q phase we have $T^{ik}_{(n);k} = -T^{ik}_{(\lambda);k} = Q^i$ and only the combined stress energy tensor is conserved. The interaction pulls down the energy density of the cosmological constant ρ_λ quite fast, while the fall of matter ρ_n occurs much more slowly than without interaction.

In the following Section 2 we mention the background results for the λ-matter interaction and the evolution of their relative energy densities in Q phase. In Section 3 we discuss a scenario where in this phase of the universe the energy transfer from the decaying cosmological constant would generate the dark matter particles with wide mass spectrum whose signatures must be observed in form of the mass dependent arcs on the CMBR power spectrum among other observable influences on structure formation [5].

2. Decay of cosmological constant through interaction in Q phase

The model for interacting cosmological constant is in part motivated by the simple consideration of the effective curvature parameter $k_{\text{eff}} = k - \lambda a^2/3$ in the Friedmann equations. The CMBR observations suggest $k_{\text{eff}} = 0$ that yields a^{-2} dependence of λ with the time dependence $\lambda \propto H^{4/3}$ in matter dominated (MD) and $\lambda \propto H$ radiation dominated (RD) universe. Several authors have discussed forms of λ dependent on H [6, 7]. With $L_\lambda/L_P \sim 10^{61}$ (where L_λ and L_P are the

λ- and the Planck length scales respectively), a plausible way to remove the inconsistency of $\lambda L_P^2 \sim 10^{-123}$ would be to find a physical mechanism that allows a swift decline of λ. We have attempted to achieve this by means of the interaction between λ and matter components.

Against the commonly prevalent view among the cosmologists, we emphasize that our local observations, by virtue of their nature, would always indicate a spatially flat universe, and it is not clear how we relate it to the global geometry [8, 9]. Any motivation to "solve the flatness problem" as claimed by inflationary theories appear to be actually silent, contrary to the common opinion, on the curvature parameter *per se* since the entire k/a^2 term is thrown out like "baby with the water" from the Friedmann equations because of inflating scale factor, and not because one 'gets' $k = 0$. (This term might haunt again in the contracting phase of a closed universe !)

In the phase I of the 3-phase model, the universe can undergo transition from RD to λ dominated universe or remain in the λ throughout this phase, depending on the initial constant value of ρ_λ. However, in the next Q phase, we have energy conservation as

$$\dot{\rho}_n + 3H\rho_n(1 + w_n) = -\dot{\rho}_\phi - 3H\rho_\phi(1 + w_\phi) = Q \tag{1}$$

with ρ_n, w_n standing for dominant background and ρ_ϕ, w_ϕ for the dark energy scalar field. The scalar fields with or without interaction have been discussed by several authors e.g. [10]. However, we consider here an interacting cosmological constant. In λCDM model $w_\phi = w_\lambda = -1$. Three alternative courses exist here. One, $w_\lambda = -1$ is held fixed throughout the evolution of the universe with $\rho_\lambda = -\dot{Q}$, as we consider in this paper. Then it is found that interaction of λ will reduce its energy density with the consequence of change in any one or more parameters of λ, c or G. Two, if all these (λ, c, G) are truly constants of time, $w_\lambda = -Q/3H\rho_\lambda - 1$ and $Q = 0$. Three, both w_λ and ρ_λ may vary like dynamical quintessence.

The scalar field dynamics is given as

$$\ddot{\phi} + 3H\dot{\phi} + V'(\phi) = -\frac{Q}{\dot{\phi}}. \tag{2}$$

The solutions of (1) provide $\rho_n \propto f a^{-3(1+w_n)}$ with $f(\lambda/M_P) = a^{Q/H\rho_n}$ as the coupling function ($M_P = (8\pi G)^{-1/2}$ is the reduced Planck mass) and keeping $Q/H\rho_n$ constant, while $\rho_\lambda \propto a^{-Q/H\rho_\lambda}$. The interaction parameter clearly depends on the variation of the coupling function as

$$Q \propto \dot{f}/a^{3(1+w_n)}. \tag{3}$$

In the absence of interaction $f = 1$ and ρ_n declines as $a^{-3(1+w_n)}$ as expected. Regardless of the background being matter or radiation dominated, $f \propto Qt^3$ for constant Q. The fast increase of coupling function with time during Q phase is mainly responsible for conversion of energy of dark cosmological constant into dark matter particles as also for slowing down the rate of fall of ρ_n in the concave-like manner. Thus, with a non-zero Q we have higher matter density at the end of Q phase than without interaction.

For a spatially flat universe with $\Omega_n^0 + \Omega_\lambda^0 = 1$, then we have

$$\frac{Q}{\dot{\phi}} \sim H_0^2 M_P^2 (1+z)^{3(1+w_n)} f'(\phi) \tag{4}$$

$$V'(\phi) = -\frac{Q}{\dot{\phi}}. \tag{5}$$

It is found that for the observed range of $w_\phi \simeq -1$, ($w = -0.969 \pm 0.061$ (stat)± 0.065 (sys)) $\dot{\phi} \approx 0$ and thus the right hand side of the (2) and (5) will be too high even for small value of Q.

This shows the important role played by the interaction even if there is no cosmological constant proper with $w_\lambda = -1$, but instead, a set of quintessence fields having $w_\phi \approx -1$. Thus the range of present observational uncertainties in determining EOS of dark energy is consistent with the concept of interaction presented in this work. The coupling function shows the time dependence of the ratio of energy densities $r = \rho_\lambda/\rho_n$ as $rf^{1+1/r} \propto a^4$ in RD and $\propto a^3$ in MD universe.

3. Diagnostics for coupling and dark matter generation

It was shown earlier that the third derivatives, and not the second derivatives, of scale factor of the universe provide effective state finders [11] for pinning down the interaction among the components as $u = \ddot{\alpha}/(aH^3)$ and $s = (u-1)/[3(q-1/2)]$ where $\alpha = \dot{a} = aH$. In our case of cosmological constant this set becomes

$$u = 1 - \frac{9}{2}\left(\frac{Q}{3H(\rho_n + \rho_\lambda)}\right) \quad (6)$$

$$s = \frac{Q}{3H\rho_\lambda}. \quad (7)$$

In terms of (7) we have the coupling function as $f = a^{3rs}$. In non-interacting phases (initial phase I and the presently ongoing phase III) no mechanism to generate dark matter can operate and so $s = 0$. However, such mechanism can clearly work in Q phase with $s \neq 0$ (though not constant). In this way the universe has greater matter density ρ_n at the end of Q phase, thus alleviating the problem of producing light nuclei like deuterium-2 and lithium-7 in the standard Friedmann-Robertson-Walker (FRW) model. In addition to it, the part of the decaying λ energy is used up in generation of dark matter. These processes must have a deep influence on the structure formation in the subsequent course of evolution.

We envisage a scenario to generate particles of dark matter (with Lorentz factor γ) each with energy

$$\xi = \int \Theta^{00} d\tau = \gamma mc^2 \quad (8)$$

where Θ^{00} is contributed by the T^{00} of λ field. For the same given ξ the more massive particles would travel more slowly than the lighter particles. Thus more massive particles would settle further away from the light cone envelopes, while the lighter particles traveling with relativistic speeds remain closer to the particle horizon. This particle mass distribution would in effect generate a multiple arc-like pattern on the Last Scattering Surface (LSS) around the recombination epoch ($z \approx 1000$) frozen on the CMBR, (and other observations [12]), each arc identifiable with a characteristic mass of definite dark matter particle (or its decay product releasing net energy $\tilde{\xi}$). Equation (3) shows the decay rate depends on the strength of interaction and as coupling grows with time as $\propto t^3$ during Q phase greater is the amount of λ energy used up to create the dark matter particles of broad mass spectrum.

It is therefore expected to have an isotropic streaming of particles, produced simultaneously (the vertex of the light cone lies at this epoch) but leaving their signatures in form of multiple arcs at LSS. For a vertex at a given epoch prior to the LSS, the arcs of smaller radii correspond to more massive particles and those of large radius towards the horizon R_H (horizon size ≈ 0.18 Mpc at $t_{\text{recombination}}$ now blown to ≈ 200 Mpc) correspond to the lighter particles. The size of an arc $L \sim R_H v/c$ for a typical particle having streaming velocity v can be related to its mass as

$$m = \xi(1 - L^2/R_H^2)^{-1/2}. \quad (9)$$

Thus, the mass density may be estimated from the vacuum considerations [13] with a suitable cut-off as

$$\Pi = \frac{\hbar}{4\pi^2}\int_0^{k_m} \omega_k k^2 dk(1 - L^2/R_H^2)^{-1/2}. \tag{10}$$

Putting the present energy scales $\sim 10^{-52}GeV^4$, $\rho_\lambda^0 \simeq 4$ GeV. This must include the uncertainty in mass given as

$$\triangle m \geq \hbar/\triangle L \tag{11}$$

where putting the maximum value of the present horizon $R_H^0 \approx 3000$ Mpc, we obtain $\triangle m \sim 10^{-32}$ eV.

We expect the intensive dark matter search experiments, like XENON100, DAMA/LIBRA, CDMS II, as well as future LUX or XMASS etc. would look for the lighter dark matter particles below ~ 10 GeV as the decay products of more massive particles, and at the same time for the signatures of decay energy $\tilde{\xi}$ released in form of γ radiation.

However, the particle production being a continuous process throughout Q phase, an overlapping of orders seems inevitable for different sets of particles produced at different epochs. In this scenario it seems that the more massive particles would be the most suitable sites for the star formation in proto-galaxies due to stronger self-gravity, and so would yield age for the youngest stars. The above conclusions of the interacting cosmological constant would be accepted only at the touchstone of such future observations.

4. Acknowledgments

The author is thankful to the University Grants Commission, New Delhi for its support for the present work through the Major Research Project vide F. No.37-431/2009 (SR).

References

[1] Riess A G et al. 2004 *Astrophys. J.* **607** 665
[2] Perlmutter S et al. 1999 *Astrophys. J.* **517** 565
[3] de Bernardis P et al. 2000 *Nature* **404** 955
[4] Verma M M 2010 *Astrophys Space Sci.* **330** 101
[5] Verma M M 2010 *Texas Symposium on Relativistic Astrophysics–TEXAS 2010, December 06-10, 2010, Heidelberg, Germany*, Proc. Science **218**
[6] Carneiro S 2006 *Int. J. Mod. Phys.* D **15** 2241
[7] Klinkhamer F R and Volovik G E 2009 *Phys. Rev.* D **79** 063527
[8] Vardanyan M, Trotta R and Silk J 2009 *Mon. Not. R. Astron. Soc.* **397** 431
[9] Springel V, Frenk C S and White S D M 2006 *Nature* **440** 1137
[10] Zimdahl W, Pavon D and Chimento L P 2001 *Phys. Lett.* B **521** 133
[11] Sahni V, Saini T D, Starobinski A A and Alam U 2003 *JETP Lett.* **77** 201
[12] Aguirre J et al. 2009 Observing the Evolution of the Universe *Preprint:* astro-ph/0903.0902
[13] Zel'dovich Ya B 1968 *Soviet Phys. Uspekhi* **11** 381

Non-Gaussian signatures arising from warm inflation driven by geometric tachyon

Anindita Bhattacharjee and Atri Deshamukhya

Department of Physics, Assam University, Silchar, Assam 788011, India.

E-mail: anindita.bhattacharjee8@gmail.com

Abstract. In a warm inflationary scenario, the initial seeds of density perturbation arise from thermal fluctuations of the inflaton field. These fluctuations in principle have Gaussian distribution. In a Gaussian distribution the density perturbation can be expressed as the two point correlation function. Thus if in an inflationary model the density perturbation is expressed as correlation function of order higher than two, these fluctuations are non-Gaussian in nature. A simple inflationary model containing single scalar field, slow roll, canonical kinetic term and vacuum initial state can produce a tiny amount of non-Gaussianity which are very small to be detected by any experiment. Non-Gaussianity can also arise in inflationary models containing multiple scalar fields. For an inflationary scenario with single scalar field, non-Gaussianity can be expressed in terms of bi-spectrum however for multi field Inflation, it is expressed in terms of trispectrum etc. In this piece of work, the warm inflationary scenario, driven by a D3 brane due to the presence of a stack of k coincident NS 5 branes is considered and the non-Gaussian effects in such an inflationary scenario has been analysed by measuring the bispectrum of the gravitational field fluctuations generated during the warm inflation in strong dissipative regime. The bi-spectrum of the Inflation is expressed in terms of the parameter $f_{\rm NL}$ and it is seen that the value of $f_{\rm NL}$ parameter lies well within the limit observed by WMAP7.

1. Introduction

Cosmic inflation refers to a period of accelerated expansion in the very early universe which was proposed more than three decades ago as a solution to the horizon and flatness problems [1]. It provides the causal interpretation of the observed anisotropy in cosmic microwave background and origin of distribution of large scale structure [2]. The scalar field which is the source of inflation is termed as inflaton. The standard inflationary scenario comprises of two varieties-cold inflation and warm inflation [3].

The cold inflationary scenario contains two regimes: slow roll and reheating epochs whereas warm inflation is an alternative form of inflation where there is no reheating epoch [4]. The basic idea of warm inflation is that the inflaton is coupled to several other scalar fields in inflationary epoch and it decays into lighter scalar fields and viscosity is generated in this process. Warm inflation ends when the universe stops inflating and it smoothly enters into the radiation dominated epoch.

In warm inflation, the density fluctuation arises from thermal fluctuations which could play a dominant roll in large scale structure formation. To a good approximation, these fluctuations have a Gaussian distribution and produces a nearly scale invariant spectrum. This means that the density perturbations have no correlation higher than second order in real and Fourier space. The simple inflationary model containing single scalar field, slow roll, canonical kinetic term and vacuum initial state can produce a tiny amount of non-Gaussianity which are too small to be detectable in present day experiments [5]. The non-Gaussianity arising in an inflationary scenario can be expressed in terms of the bi-spectrum [6, 7], trispectrum etc [8] and they can be measured in the distribution of cosmic microwave background fluctuations [9, 10]. Other scenarios such as the multiple-field inflation models can also give ideas about non-Gaussianity measurements [11].

In the last decade, there have been many attempts to describe successful models of inflation driven by scalar fields motivated from String Theory. Tachyon field associated with unstable D-brane is one such scalar field which may be responsible for driving inflation in the early phase of universe [12, 13]. When one considers a D3 brane in the background of k coincident NS 5 brane, the system gives rise to a unstable tachyonic field termed as *Geometric Tachyon*. We have investigated the warm inflationary scenario driven by this geometric tachyonic field and found that non-Gaussian signals predicted from this model lies well within the range predicted by WMAP7 [14, 15].

In section 2, the non-Gaussianity measurement for an warm inflationary scenario is discussed and in section 3, the non-Gaussianity for geometric tachyon case has been analysed.

2. Non-Gaussianity and warm inflation

The equation of motion for the zero mode of the scalar field ϕ in general has the form

$$\frac{d\phi}{dt} = -\frac{1}{\Gamma}\frac{dV}{d\phi} \tag{1}$$

In order to study the non-Gaussianity, one needs to obtain the three point correlation function of the density perturbation or the bispectrum. For this reason, we need to study the fluctuation of the tachyon field $\delta\phi(x,t)$. It is assumed that they are small and the full tachyon field can be expressed as $\phi(x,t) = \phi_0(t) + \delta\phi(x,t)$, where ϕ_0 is the homogeneous background field. Now for measuring the bi-spectrum, the equation of motion needs to be considered up to second order fluctuations

$$\delta\phi = \delta_1\phi + \delta_2\phi \tag{2}$$

It is often very useful to express these quantities in momentum space. From equation (1), the equations of motion for first and second order perturbation can be found out [16], the solution of which are of the form

$$\delta_1\phi(k,t) = A(k,t-t_{n-1})\int_{t_{n-1}}^{t} dt'\frac{\eta(k,t')}{\Gamma}A^{-1}(k,t'-t_{n-1}) + A(k,t-t_{n-1})\delta\phi_1(ke^{-H(t_n-t_{n-1})},t_{n-1}) \tag{3}$$

and

$$\begin{aligned}\delta_2\phi(k,t) &= A(k,t-t_{n-1})\int_{t_{n-1}}^{t} dt' B(k,t')\left[\frac{dp^3}{(2\pi)^3}\delta_1\phi(p,t')\delta_1\phi(k-p,t')\right]A^{-1}(k,t'-t_{n-1}) \\ &+ A(k,t-t_{n-1})\delta_2\phi(ke^{-H(t_n-t_{n-1})},t_{n-1})\end{aligned} \tag{4}$$

where,

$$A(k,t) = \exp\left[-\int_{t_0}^{t}\left(\frac{k^2V(\phi_0(t'))}{\Gamma} + \frac{V''(\phi_0(t'))}{\Gamma}\right)dt'\right] \tag{5}$$

$$B(k,t) = -\frac{2V'(\phi_0(t))k^2 + V'''(\phi_0(t))}{\Gamma} \tag{6}$$

In both these equations, the second term symbolises the memory terms within the Hubble time which leads the mechanism of freeze out [16]. From the definition of freeze out, for $k \geq k_F$ the memory terms damp away within a Hubble time and for $k \leq k_F$ they do not damp. To satisfy this fact, a quantity called the freeze out momentum k_F can be defined with the condition

$$\frac{V(\phi_0)k^2 + V''(\phi_0)}{\Gamma H} > 1 \tag{7}$$

For warm inflation, $V''(\phi_0) < \Gamma H$ and hence k_F can be defined as

$$k_F = \sqrt{\frac{\Gamma\, H}{V(\phi_0)}} \tag{8}$$

By using the solutions (3) and (4), the three-point correlation function of the fluctuations before the end of inflation (when $t = t_{60}$) can be computed. As this scenario designates single field inflationary mechanism, so, the three-point correlation function defined in this picture is called the bi-spectrum. In the freeze-out region ($k > k_F$) using the approximation $A \simeq 1$ and $B(k,t) \approx B(k_F, t_F)$, the three point correlation function or the bi-spectrum can be computed and from this expression one can determine the $f_{\rm NL}$ parameter as

$$f_{\rm NL} = -\frac{5}{3}\left(\frac{\dot{\phi}}{H}\right)\left[\frac{1}{H}\ln(\frac{k_F}{H})\right]\left(\frac{V'''(\phi_0(t_F)) + 2k_F^2 V'(\phi_0(t_F))}{\Gamma}\right) \tag{9}$$

As $f_{\rm NL}$ is the measure of non-Gaussianity and so, for an inflationary model, one needs to find out the expression of $f_{\rm NL}$ parameter, the value of which decides whether non-Gaussianity arises in this model or not.

3. Non-Gaussianity arising in geometric tachyon model

When the one dimensional motion of a Dp-brane inside a ring of NS 5-branes is considered, the radiation field becomes tachyonic [15] and this arrangement is termed as Geometric Tachyon. The potential under this configuration of the geometric tachyon field is given by

$$V(\phi) = V_0 \cos\left(\frac{\phi}{\sqrt{k\, l_s^2}}\right), \qquad V_0 = \frac{\tau_3 R}{\sqrt{k\, l_s^2}}. \tag{10}$$

with the condition

$$\sqrt{k\, l_s} \geq R \tag{11}$$

where k is the number of NS5 branes in the ring of radius R and l_s is the string length scale.

We restrict ourselves in the region of high-dissipation where, $\Gamma \gg 3HV$ i.e. $r = \gamma/(3HV) \gg 1$. Using the slow-roll approximation, the number of e-foldings , $N = \ln a$, a being the scale factor then becomes

$$N = \int H dt = -\int \frac{3H^2 r V(\phi)}{V'(\phi)} d\phi = p[y - y_f]. \tag{12}$$

where $y = \cos(\phi/\sqrt{kl_s^2})$, $x = (\phi/\sqrt{kl_s^2})$ and p is a dimensionless parameter introduced via $p = V_0/(4d)$ and d is another parameter in this scenario, which is a combination of the radiation temperature T_γ and the Stephan-Boltzmann constant σ. From the condition of the end of inflation and also, using the fact that one needs $N \sim 60$ before the end of inflation, we can get an expression of y by a suitable choice of the parameter p. By choosing $p = 10^2$, the initial value of y comes out to be $y = 0.60500025$.

Journal of Physics: Conference Series **481** (2014) 012012 doi:10.1088/1742-6596/481/1/012012

For this case, the potential in terms of x is written as $V(x) = \cos(x) = y$ and so, $f_{\rm NL}$ in strong dissipative regime can be written as

$$f_{\rm NL} = \frac{5}{3p^2 V(x)^{\frac{3}{2}} V'(x)} \frac{4V(x)V'(x)^3 + 7V(x)^2 V'(x)V''(x) + V(x)^3 V'''(x)}{V(x)^{\frac{3}{2}} V'(x)^2} \tag{13}$$

$$+ 2\sqrt{V(x)}V'(x)p) \log\left[\frac{\sqrt{3}\sqrt{V'(x)^2}}{2\sqrt{d}}\right]$$

$$= \frac{10\left[\frac{2}{y^2} + \frac{p}{y} - \frac{4}{1-y^2}\right] \log\left[\frac{\sqrt{3}\sqrt{1-y^2}}{2\sqrt{d}}\right]}{3p^2} \tag{14}$$

When $p = 100$, $x = .60500025$, $d = 10^{-12} M_p^4$, $f_{\rm NL}$ becomes 0.736917 which is well within the bound observed by WMAP7.

4. Conclusion

In this work, it is predicted that non-Gaussian signals in the density spectrum of the scalar field can arise in the warm tachyonic inflationary scenario driven by geometric tachyon. As non-Gaussianity is generally measured in terms of the $f_{\rm NL}$ parameter of the bi-spectrum, we have estimated the value of $f_{\rm NL}$ in this model. It is seen that its value lies well within the limit predicted by WMAP7 for the range of model parameters.

References

[1] A. H. Guth, Phys. Rev. D **23** (1981) 347-352.
[2] A. J. Albrecht and P. J. Steinhardt, Phys. Rev. Lett. **48**, 1220-1223 (1982). A.Linde, Particle Physics and inflationary cosmology (Gordon and Breach, New York, 1990).
[3] A. Berera, Phys. Rev. Lett. **75**, 3218 (1995) [arXiv:astro-ph/9509049].
[4] L. M. H.Hall,I. G.Moss and A. Berera, Phys. Rev. D. **69**, 083525 (2004); A.Berera,Nucl.Phys B 585,666(2000).
[5] A. Gangui,F. Lucchin,S. Matarrese and S. Mollerach, ApJ,**430**, 447 (1994).
[6] A.F. Heavens, MNRAS,**299**, 805 (1998).
[7] P. Ferreira,J. Magueijo and K. Gorski, ApJ Lett,**503**, 1 (1998).
[8] M. Kunz,A. J. Banday,P. G. Castro,P. G.,P. G.Ferreira and K. M.Grski, ApJ Lett,**563**, L99 (2001).
[9] L. Verde,L. Wang,A. F. Heavens and M. Kamionkowski, Phys. Rev. D **61**, 063504 (2000).
[10] L. Wang,M. Kamionkowski, ApJ Lett,**503**, 1 (1998).
[11] S. Gupta A. Berera, A. F. Heavens and S. Matarrese, Phys. Rev. D **66**, 043510 (2002) [arXiv:astro-ph/0205125V2].
[12] A. Sen, Mod.Phys.Lett.A,**17**, 1797 2002).
[13] M. Sami,P. Chingangbam,T. Qureshi Phys. Rev. D,**66**, 043530 (2002).
[14] Ashok Das,Sudhakar Panda,Shibaji Roy, Phys. Rev. D78,061901(2008).
[15] Sudhakar Panda,M. Sami,Shinji Tsujikawa, Phys. Rev. D73,023515(2006).
[16] M. R. Setare and V. Kamali, JCAP,**08**,034 (2012)[arXiv:1204.2156[hep-th]]

Study of IGM through high energy radiation from blazar

Jayashri Medhi, H L Duorah, A G Barua, and K Duorah
Department of Physics, Gauhati University, Guwahati 781014, Assam, India.
E-mail: jayashri.medhi@rediffmail.com

Abstract. The high energy gamma rays from blazar affects the intergalactic medium (IGM) to large distances at different redshifts. The blazar radiation has been taken as the result of synchrotron and Inverse Compton scattering of electrons. It is found that the intergalactic medium is clumpy. Our estimated values lie within the suggested limit of $\Omega_{IGM} \cong 0.03$ at redshift $z = 3$.

1. Introduction

Blazars are subclass of active galactic nuclei (AGN) where high energy emission makes them ideal candidates for gamma ray observations. Blazars jointly comprise flat spectrum radio quasars (FSRQs) with broad, strong emission lines and BL Lacertae ("BL Lac") objects, which show no evidence for emission lines [1, 2]. Blazars are characterised by a relativistic plasma jet oriented at a small angle with respect to the observer's line of sight. They are broadband sources often accompanied by prominent emission of electromagnetic radiation throughout the entire electromagnetic spectrum. Radio to optical emission in the blazar, s spectrum is produced by synchrotron radiation of relativistic electrons while gamma rays are most naturally to be produced by the inverse Compton scattering of soft photons by the same electron population [3, 4]. In this paper we present synchrotron and Inverse Compton mechanism for γ-ray production in blazar. As very high energy gamma rays (VHEGRs) propagate through the Universe, they interact with the soft photons that comprise the extragalactic background light (EBL) and heats the intergalactic medium (IGM) efficiently by producing ultra- relativistic pairs. Due to blazar heating the temperature of the IGM increases to 10^4 K- 10^6 K [5].

2. Synchrotron emission

Synchrotron radiation is produced when the energetic electrons spiraling in a magnetic field of blazar jets. The photons emitted as synchrotron radiation have characteristic frequency given by

$$\omega_c = \frac{3}{2}\gamma^2 \left(\frac{eH}{mc}\right), \tag{1}$$

where e and m are the charge and mass of the electron, H is the strength of the magnetic field, γ is the Lorentz factor and c is the speed of light. If we take a power-law spectrum of the form $I_e(E) = K_e E^{-\Gamma}$ for electron energy $E = \gamma mc^2$,the differential flux given by [6] is

$$I_{\text{sync}}(E_\gamma) = 9.46 \times 10^{-10} \left(8.15 \times 10^{-6}\right)^{\Gamma} a(\Gamma) L \times K_e H^{\frac{(\Gamma+1)}{2}} E^{-\frac{(\Gamma+1)}{2}}, \tag{2}$$

Table 1. Intensities and fluxes for Synchrotron radiation in the energy range from 10^{-3} eV to 10^{3} eV

Energy (eV)	$\mathrm{I}(E_\gamma)\,\mathrm{cm^{-2}s^{-1}sr^{-1}GeV^{-1}}$	Flux ($\times 10^{-21}\mathrm{ergs\,cm^{-2}s^{-1}sr^{-1}Hz^{-1}}$)
10^{-3}	8.13×10^{18}	66
10^{-2}	14.6×10^{16}	11.8
10^{-1}	26×10^{13}	2.1
10^{1}	8.13×10^{11}	0.66
10^{2}	14×10^{9}	0.012
10^{3}	26×10^{7}	0.002

Where $a(\Gamma)$ is a numerical coefficient with values $a(2.5) = 0.85, a(3) = 0.074$ and $a(4) = 0.072$. Taking $\Gamma = 2.5$ and the magnetic field in emission region $H = 0.1\,\mathrm{G}$ we have calculated the differential flux for the energy range from 10^{-3} eV to 10^{3} eV.

3. Inverse Compton emission

Inverse Compton scattering is generally thought to be responsible for the production of high energy gamma rays in blazars. Several suggestions have been made for the origin of the target photons : optical/uv photons from an accretion disk [7], optical photons from the broad line region and infrared emission of dusty torus [8, 9] can produce the target photons. This radiation mechanism is usually called External Comptonization (EC). Synchrotron photons produced in the jet can also act as target photons for inverse Compton scattering, a process referred to as Synchrotron Self Compton (SSC) scattering [3, 10]. Here we have assumed SSC model for gamma ray emission from blazar. Application of this model to the observations of TeV γ-ray emission from Mrk 421 has been recently tested by [11]. Based on the detected variability time scales of TeV γ-ray emission, observations of coincident flares in X-rays and TeV γ-rays and the observed multiwavelength photon spectrum of Mrk 421, the constraints has been placed on the allowed parameter space (magnetic field in the emission region and its Doppler factor) for the homogeneous SSC model. For the 1 day flare, the magnetic field in the blob has to be limited to the range $\sim 0.025 \div 0.15$ G and the corresponding Doppler factors to the range $\sim 20.5 \div 10.7$. For the 15 min flare these limits are following: $\sim 0.4 \div 1.3$ G and $\sim 37.6 \div 24$.
For homogenous, isotropic electron intensity distribution, the differential γ-ray intensity produced by Inverse Compton scattering of relativistic electrons suggested by [12] is

$$I(E_\gamma) = \frac{1}{2} L \rho_{\mathrm{ph}} \sigma_T \left(mc^2\right)^{1-\Gamma} (\bar{\varepsilon})^{\frac{(\Gamma-3)}{2}} K_e E_\gamma^{-\frac{1}{2}(\Gamma+1)} f(\Gamma), \tag{3}$$

Where $f(\Gamma) \cong 1$ for Γ in the range 1 to 2. We assume $L = 7 \times 10^{22}$ cm, radiation density $\rho_{\mathrm{ph}} = 4.75 \times 10^{-8}$ GeV cm^{-3} , $\sigma_T = 6.65 \times 10^{-25}$ cm^2, Mean photon density $\bar{\varepsilon} = 2.3 \times 10^{-5}$ GeV and $\Gamma = 2$.

4. Impact of blazar on the intergalactic medium (IGM)

When the high energy gamma rays emitted from the blazar passes through the IGM they interact with the optical light produced by the galaxies and transforming it into the elementary particles electrons and positrons. These very high energy gamma ray photons are efficiently converted

Journal of Physics: Conference Series **481** (2014) 012013 doi:10.1088/1742-6596/481/1/012013

Table 2. Intensities and fluxes due to Inverse Compton Scattering in the energy range 10 GeV-10 TeV

Energy (GeV)	$I(E_\gamma)\,\mathrm{cm^{-2}s^{-1}sr^{-1}GeV^{-1}}$	Flux ($\times 10^{-31}\mathrm{ergs\,cm^{-2}s^{-1}sr^{-1}Hz^{-1}}$)
10	1.5×10^{-5}	12
100	4.6×10^{-7}	3.7
1,000	1.5×10^{-8}	1.2
2,000	0.53×10^{-8}	0.8
4,000	0.19×10^{-8}	0.6
6,000	0.1×10^{-8}	0.48
8,000	0.066×10^{-8}	0.4
10,000	0.047×10^{-8}	0.38

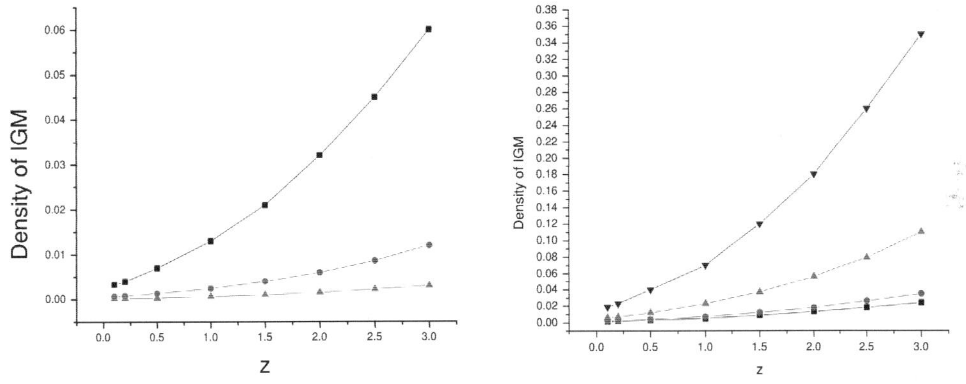

Figure 1. IGM density vs. redshift in case of Synchrotron radiation(left)/IGM density vs. redshift in case of Inverse Compton radiation (right)

into heat in IGM. Blazar increases the temperature of the IGM above 10^4 K [4]. At temperature 10^4 K,

$$\Omega_{IGM} < 0.4 I_{21} \Omega^{\frac{1}{2}} h^{-\frac{3}{2}} (1+z)^{\frac{9}{2}}\,, \qquad (4)$$

where I_{21} is the ionizing flux and is measured in the units of 10^{-21} ergs cm^{-2} s^{-1} Hz^{-1} sr^{-1} [13]. Here we have calculated the value of Ω_{IGM} for different intensities of radiation emitted by the blazar at redshift $Z = 3, 2.5, 2, 1.5, 1, 0.5, 0.2$, and 0.1.

5. Result and discussion

In this work, we have explored the γ-ray emission mechanisms due to synchrotron and Inverse Compton scattering at different energy range. From the graph, we may say that density of IGM is different at different Z showing that the intergalactic medium is basically clumpy. This clumpiness may be due to the radiation pressure driving the intergalactic matter away in space. It is seen that density of IGM increases at higher redshift and decreases as ionizing flux decreases. The straight line parallel to x-axis shows the upper limit of $= 0.03$. This gives us an idea about the region of same density of the intergalactic medium for different ionizing flux of blazar radiation at different redshift.

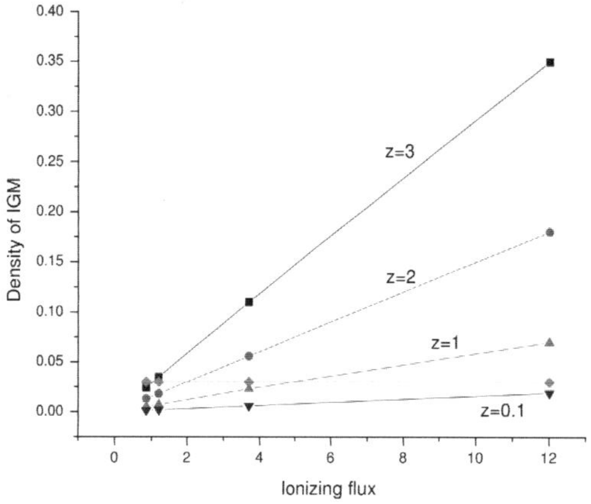

Figure 2. Density of IGM vs ionizing flux at redshift $z = 3, 2, 1,$ and 0.1 in case of IC scattering. Ionizing flux in the units of $10^{-31}\mathrm{ergs\,cm^{-2}s^{-1}Hz^{-1}sr^{-1}}$

References

[1] Angle J R P , Stockman H S 1980 *Ann. Rev. of Astron. Ap.* 18 321-361
[2] Urry C M , Padovani P 1995 http://arxiv.org/abs/astro-ph/9506063
[3] Maraschi L, Ghisellini G and Celotti A 1992 *ApJ* 397 L5.
[4] Dermer C D and Schlickeiser R 1993 *The Astrophysical Journal* 416, 458
[5] Broderick A E , Chang P, and Pfrommer C 2012 ApJ in print arXiv:1106.5494
[6] Ginzburg V L and Syrovatskii S I 1965 *Ann. Rev. Astron. Ap.* 3 297-350
[7] Dermer C D, Schlickeiser R and Mastichiadis A 1992 *Astron. Ap.* 256 (2)27-30.
[8] Sikora M , Begelman M C and Rees M J 1994 *The Astrophysical Journal* 421 153
[9] Blazejowaski M, Sikora M, Moderski R and Madejski G M 2000 *The Astrophysical Journal* 545 107
[10] Bloom S D and Marscher A P 1996 *Astrophysical Journal* 461 657-663
[11] Bednarek W 1997 arXiv: astro-ph/9711189
[12] Felten J E 1965 (Ph.D. thesis, Cornell University)
[13] Coles P and Lucchin F Cosmology The Origin and Evolution of Cosmic structure second edition 20 430

Thermodynamics of BTZ black hole and entanglement entropy

Dharm Veer Singh and Sanjay Siwach

Department of Physics,Banaras Hindu University, Varanasi 221005, India.

E-mail: veerdsingh@gmail.com,sksiwach@gmail.com

Abstract. The BTZ black Hole is (2+1) dimensional black hole solution asymptotic to anti-de-Sitter space-time. We study the discretized quantum scalar fields in background of non-rotating BTZ black hole space-time and construct the entanglement thermodynamics for massless scalar field. The behavior of the entanglement energy is understood by red shift factor caused by the curved background. The entanglement thermodynamics is compared with the black hole thermodynamics.

1. Introduction

Black hole is region of space time where gravitational pull is so strong that even light can not escape. The event horizon is defined as the null surface in space time. The total area of the event horizon never decreases in any physical process. This is analogous to entropy in thermodynamics, which never decreases in a physical process. Thus the black holes can be assigned entropy proportional to the area of black hole horizon. One can ask the question about microscopic degrees of freedom responsible for entropy as nothing comes out of black holes. The quantum mechanical treatment predicts that the black holes radiates and the radiation carries no information as it is thermal in nature.

In this paper we carry the investigation of entanglement thermodynamics of BTZ black hole in three dimensional space-time with negative cosmological constant [1, 2]. This solution can be embedded in string theory and a microscopic counting of degrees of freedom responsible for black hole entropy is also known [3]. The near horizon geometry of BTZ black hole is anti-de-Sitter space time and the machinery of AdS/CFT correspondence can also be applied. Thus BTZ black hole can be used as a laboratory to study the quantum aspects of black holes. There are several attempts to understand the microscopic origin of black hole entropy and its relation with horizon area [2–14].

In this paper, we investigate the entanglement thermodynamics of scalar fields in the background of BTZ black hole space-time. The entanglement entropy $S_{\rm ent}$ is always proportional to the area A of the event horizon of the black hole as expected, $S_{\rm ent} = C_S\, A/a$. Here 'a' is cut-off length and C_s is model dependent coefficient, which we have calculated numerically in this paper. Entanglement entropy $S_{\rm ent}$ is independent of the position of an observer once a quantum state is fixed [7, 10, 14]. We also calculate the entanglement energy of the BTZ black hole. The behavior of the entanglement energy is understood by the redshift factor caused by curved background. Because of the gravitational redshift, these energies are related as $E = \sqrt{-g_{tt}}\, E_{\rm ent}$.

2. Entanglement entropy of scalar field in non-rotating BTZ black hole

The BTZ black hole is a solution of Einstein's gravity in (2+1) dimensions with a negative cosmological constant ($\frac{1}{l^2}$). The metric of non rotating BTZ black hole can be written as [1];

$$ds^2 = -(-M + r^2/l^2)dt^2 + \frac{1}{(-M + r^2/l^2)}dr^2 + r^2 d\phi^2. \tag{1}$$

The horizon of the black hole is located at $r_+ = \sqrt{M}\,l$. The proper length from the horizon is defined as, $\rho = \int_{r_+}^{r}(r'^2/l^2 - M)^{-1/2}\,dr' = l\log[(\sqrt{u^2+M}+u)/r_+]$, where we have defined $u = \sqrt{r^2/l^2 - M}$. The metric can be rewritten in term of proper distance as:

$$ds^2 = -u^2 dt^2 + d\rho^2 + r_+^2\left(\frac{u^2}{M} + 1\right) d\phi^2. \tag{2}$$

Let us consider a massless scalar field in BTZ black hole space-time. Using the separation of variable as appropriate for the cylindrical symmetry of the system; $\Phi(t,\rho,\phi) = \sum_m \Phi_m(t,\rho)\,e^{im\phi}$, and defining a new variable, $\psi_m(t,\rho) = \left[\left\{r_+\sqrt{(u^2/M)+1}\right\}/u\right]^{1/2}\Phi_m(t,\rho)$, the Hamiltonian of the system can be written as,

$$H = \frac{1}{2}\int d\rho\,\pi_m^2(\rho) + \frac{1}{2}\int d\rho\,d\rho'\,\psi_m(\rho)\,V_m(\rho,\rho')\,\psi_m(\rho'),$$

where

$$\psi_m(\rho)V(\rho,\rho')\psi_m(\rho') = \left[u\sqrt{u^2/M+1}\left\{\partial_\rho\left(\sqrt{\frac{u}{\sqrt{(u^2/M+1)}}}\psi_m\right)\right\}^2 + m^2\frac{u^2}{r_+^2\,(u^2/M+1)}\psi_m^2\right]. \tag{3}$$

The system can be discretized as $\rho \to (A-1/2)a$ and $\delta(\rho-\rho') \to \delta_{AB}/a$, where A,B=1,2 ....N and a is cut-off length. The corresponding Hamiltonian of the discretized system can be obtained by the replacement: $\psi_m(\rho) \to q^A$, $\pi_m(\rho) \to p_A/a$, $V(\rho,\rho') \to V_{AB}/a^2$.

The discretized Hamiltonian resembles to the set of N coupled harmonic oscillators [5, 6],

$$H = \sum_{A,B=1}^{N}\left[\frac{1}{2a}\delta^{AB}p_A\,p_B + \frac{1}{2}V_{AB}\,q_m^A\,q^B\right] \tag{4}$$

To calculate the entanglement entropy, we divide the system into two sub-systems and define,

$$(W)_{AB} = \sqrt{a}\,V_{AB} = \begin{pmatrix} A_{ab} & B_{a\beta} \\ (B^T)_{ab} & D^{(2)}_{\alpha\beta} \end{pmatrix},$$

and the indices 'a,b..' and 'α,β...' refer to two subsystems.

The entanglement entropy of the system is given by $S = -tr[\rho\log\rho]$, where ρ is the density matrix. The density matrix of the system is defined as,

$$\rho(q,q') = \int\prod_{i=1}^{N}dq_i\psi(q_1,......q_n;q_{n+1},.......q_N)\psi^\star(q_1,......q_n;q_{n+1},.......q_N). \tag{5}$$

The corresponding density matrix can be calculated as,

$$\rho_0(q,q') = \exp\left[-(q^TD'q + q'^TD'q')/2 + q^T(D-D')q\right] \tag{6}$$

where $D' = D - \frac{1}{2}B^TA^{-1}B$.

The entanglement entropy can be written as $S_{\text{ent}} = \sum_{i=1}^{N-n_B}S_i$, where $S_i = -\mu_i(1-\mu_i)^{-1}\ln\mu_i - \ln(1-\mu_i)$ and $\mu_i := \lambda_i^{-1}(\sqrt{1+\lambda_i}-1)^2$. The λ_i are the eigenvalues of the W_{AB} matrix. Using these relations, we evaluate the entanglement entropy of the system.

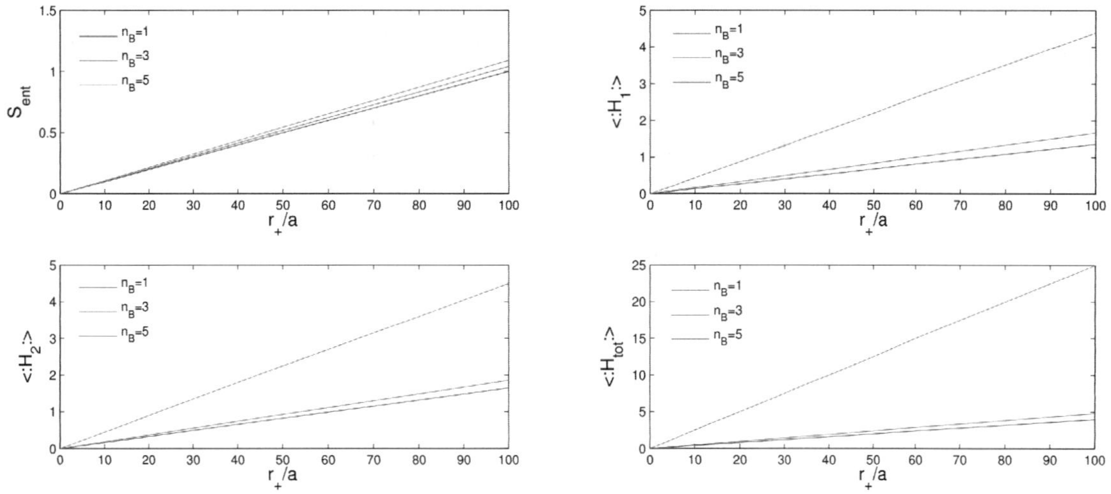

Figure 1. The numerical calculation for $S_{\rm ent}, H_1$ (upper panel), $H_2, H_{\rm tot}$ (lower panel) of the scalar field in BTZ space-time, for n_B =1,3,5 is shown as a functions of r_+/a with $N = 100$ and $m = 0$.

3. Numerical estimation

3.1. Entanglement entropy

We are interested in the numerical estimation of the entanglement entropy of the scalar fields in the background of BTZ black hole for the above mentioned system. The result for the entanglement entropy of the discretized system is; $S_{\rm ent} \simeq 0.01\,(r_+/a)$, where the proportionality constant is evaluated numerically for $n_B = 1$. The results are shown in Fig.1 for different values of n_B. The entropy is proportional to r_+, which in turn is proportional to horizon area.

3.2. Entanglement energy

We also calculate the entanglement energy of the scalar fields in the background of BTZ black hole. For the subsystem '1' the entanglement energy is given by,

$$\langle : H_1 : \rangle = \frac{1}{4a}\left[aV^{(1)}_{ab}(\tilde{A})^{ab} + A_{ab}\delta^{ab} - 2w^{(1)}_{ab}\delta^{ab} \right] \tag{7}$$

Using this, we estimate the numerical value of the entanglement energy. $\langle : H_1 : \rangle \simeq 0.014\left(\frac{r_+}{a}\right)$, The results are shown in Fig.1 for different values of n_B.

Similarly for the subsystem '2' the entanglement energy is given by,

$$\langle : H_2 : \rangle = \frac{1}{4a}\left[aV^{(2)}_{\alpha\beta}(\tilde{D})^{\alpha\beta} + D_{\alpha\beta}\delta^{ab} - 2w^{(1)}_{\alpha\beta}\delta^{\alpha\beta} \right] \tag{8}$$

Using this, we estimate the numerical value of the entanglement energy.$\langle : H_2 : \rangle \simeq 0.019\left(\frac{r_+}{a}\right)$. The results are shown in Fig.1 for different values of n_B.

We also estimate the total energy of the system ,

$$\langle : H_{tot} : \rangle_{\rho'} = \frac{1}{4a}Tr[aVM^{-1} - W] = \frac{1}{4}Tr[V(M^{-1} - W^{-1})] = -\frac{1}{2}[V^T_{int}\tilde{B}], \tag{9}$$

which is given by;$\langle : H_{tot} : \rangle_{\rho'} \simeq .045\left(\frac{r_+}{a}\right)$, The results for total energy are shown in Fig.1 for different values of n_B.

The entanglement entropy and energy of the BTZ black hole looks like as $S_{\text{ent}} \propto A$ and $E_{\text{ent}} \propto A$, where $A = 2\pi r_+$ is area of the black hole and $r_+ = \sqrt{Ml}$. The corresponding entanglement entropy and energy of the thermal AdS space-time are given as, $S_{\text{ent}} \propto A$ and $E_{\text{ent}} \propto A^2$, where A is area of the slice of radius r. Thus the discrepancy between entanglement energy of the BTZ black hole and thermal AdS space time. This discrepancy can be removed by consideration of the redshift factor.

The entropy of black hole has the same behavior as that measured by the observer at infinity, since the degree of freedom are independent of n_B. The results for the entanglement energy is dependent on n_B and one can try to understand this is in terms of red shift of energy $\sqrt{-g_{tt}}E_{\text{ent}}$ in curved spacetime [7], as measured by an observer who is at proper distance $n_B\, a$ from the horizon. The entanglement energy is redshifted by the factor $\rho\, r_+$, where $\sqrt{-g_{tt}} = \rho\, r_+$, and ρ is the proper distance from the horizon correspond to stationary observer. The gravitational effect restores the agreement between entanglement thermodynamics and black hole thermodynamics.

4. Conclusion

In this paper, we have calculated the entanglement entropy and energy of quantum scalar fields in BTZ black hole space-time. The model is very similar to a bunch of harmonic oscillators for free fields in curved space-time but seem to capture area law of black hole thermodynamic. The results can be compared with corresponding quantum field entropy in Schwarzschild black hole space-times in four dimensions [7]. The discrepancy between BTZ Black hole thermodynamics and thermal AdS space-time has been removed by gravitational redshift factor. It would be interesting to generalize these results for Fermionic degrees of freedom.

Acknowledgment

The work of Dharm Veer Singh is supported by Rajiv Gandhi National Fellowship Scheme (U.G.C.) of Government of India.

Reference

[1] Maximo Banados, Claudio Teitelboim and Jorge Zanelli,"The Black Hole in Three Dimensional Spacetime" Phy. Rev. Lett. **69**(1992) 1849-1851.

[2] S. Carlip, "The (2+1)-Dimensional Black Hole" Classical and Quantum Gravity **12**(1995) 2853-2880, `arXiv:gr- qc/9506079v1`.

[3] A. Strominger and C. Vafa, "Microscopic origin of the Bekenstein-Hawking entropy," Phys. Lett. B **379** (1996) 99, `arXiv:hep-th/9601029`.

[4] Sung-Won Kim,Won T. Kim, Young-Jai Park, and Hyeonjoon Shin,"Entropy of BTZ black hole in(2+1) dimensions" Phys. Lett B **392**,311-318(1997).

[5] Luca Bombelli, Rabinder K. Koul, Joohan Lee and Rafael D. Sorkin,"Quantum Source of entropy for Black Hole" Phy. Rev. D **34**, 373 (1986).

[6] Mark Srednicki,"The Entropy and Area" Phy. Rev. Lett. **71** 666 (1993).

[7] Shinji Mukohyama, Masafumi Seriu and Hideo kodama,"Thermodynamics of entanglement in Schwarzschild Space time" Phy. Rev. D **58**, 064001 (1998).

[8] Shinji Mukohyama,"Comment on entanglement entropy" Phy. Rev. D **58**, 104023 (1998).

[9] Shinji Mukohyama, Masafumi Seriu, and Hideo Kodama, "Can the entanglement entropy be the origin of black-hole entropy? "Phys. Rev. D **55**,7666(1997).

[10] H.Casini and M.Huerta, "Entanglement Entropy in free quantum theory"`arXiv:hep-th/09052562v3`.

[11] V. Frolov, "Why the entropy of Black hole is A/4" Phy. Rev. Lett. **74** 3319 (1995).

[12] Danny Birmingham, Ivo Sachs and Siddhartha Sen,"Exact results of BTZ black hole" `arXiv:hep th/0102155v2`.

[13] M.R.Setare, "Non-rotating BTZ black hole area spectrum from quasi-normal modes" Classical and Quantum Gravity **21** 1453-1457(2004).

[14] Dharm Veer Singh and Sanjay Siwach,"Scalar Fields in BTZ Black Hole Spacetime and Entanglement Entropy," arXiv:hep-th/1106.1005.

Thermodynamic origin of interaction in the cosmological tachyonic field

S D Pathak[1] and M M Verma[2]

Department of Physics, University of Lucknow, Lucknow 226007, India.

E-mail: [1]prince.pathak19@gmail.com, [2]sunilmmv@yahoo.com

Abstract. We propose a model with thermodynamic origin of interaction among the components of tachyonic scalar field in the universe. The components are taken to exist in the state of non-equilibrium initially but undergoing a transition towards the equilibrium state. We show that this transition requires interaction among the components. During such transition phase, the transfer of energy among the components is governed by the second law of thermodynamics with local violation of conservation of energy for individual components. It is further shown that in our proposed mechanism the interaction itself generates an increase of entropy in an evolving universe, and thus it might indicate a possible solution to the well-known entropy problem.

1. Introduction

Our universe is currently in a phase of accelerated expansion as interpreted through the cosmological observations over the past decade such as SNe type Ia [1, 2], Cosmic Microwave Background Radiation(CMBR) [3], Baryon Acoustic Oscillations(BAO) [4] in galaxy surveys etc. It is understood that nearly 73% of the total content in the universe is composed of an exotic component with negative pressure which, often termed as dark energy, is held responsible for this accelerated expansion at the present epoch. The Lagrangian $L = -V(\phi)\sqrt{1 - \partial_i\phi\partial^i\phi}$ for the proposed tachyonic scalar field ϕ taken from [5] (arising from string theory) provides with the corresponding action as

$$\mathcal{A} = \int d^4x\sqrt{-g}\left(\frac{R}{16\pi G} - V(\phi)\sqrt{1 - \partial^i\phi\partial_i\phi}\right) \tag{1}$$

and the energy-momentum tensor $T^{ik} = \frac{\partial L}{\partial(\partial_i\phi)}\partial^k\phi - g^{ik}L$ yields the energy density and pressure (for spatially homogeneous field approximation) as

$$\rho = \frac{V(\phi)}{\sqrt{1 - \dot{\phi}^2}}; \qquad P = -V(\phi)\sqrt{1 - \dot{\phi}^2} \tag{2}$$

where an overdot denotes the derivative with respect to time (as in the rest of this paper). Here, we consider ρ and P as composed of two components [6] which fit in so well that if one component is matter with $w_m = 0$, then the other one turns out to be the cosmological constant with $w_\lambda = -1$. Thus, we obtain these components from (2), respectively, as

$$\rho_m = \frac{V(\phi)\dot{\phi}^2}{\sqrt{1-\dot{\phi}^2}}; \qquad P_m = 0 \tag{3}$$

and

$$\rho_\lambda = V(\phi)\sqrt{1-\dot{\phi}^2}; \quad P_\lambda = -V(\phi)\sqrt{1-\dot{\phi}^2}. \tag{4}$$

2. Interaction between components

Next, we study the mutual interaction of components of tachyonic scalar field. The interacting dark energy models have been recently proposed by several authors [7–10]. Under these circumstances, the cosmological constant is no longer a true constant and the rate of its decline is ascertained by the complementary evolution of dark matter energy densities. It is obvious that during interaction the overall conservation of energy is kept intact. The components are in non-equilibrium (thermal, mechanical or else) and to achieve the equilibrium the components interact mutually. The transfer of energy is governed by the second law of thermodynamics. If the components are in equilibrium then due to interaction perturbation tries to restore or achieve new stable state (Le Châtelier-Braun principle) [11, 12]. The individual equations of energy conservation for dark matter ($w_m = 0$) and the cosmological constant ($w_\lambda = -1$) are respectively given as

$$\dot{\rho_m} = -3H\rho_m + Q; \quad \dot{\rho_\lambda} = -Q. \tag{5}$$

Considering the dimensional compatibility it is natural choice that the interaction strength Q should be function of the Hubble parameter and energy density. Due to the lack of information regarding the exact nature of dark matter and dark energy (as the cosmological constant or else) we could not fixed the exact form of interaction strength. Thus, with this motivation we present the form of interaction term, heuristically, as the function of time rate of change in energy density. Thus, the simple assumption about the interaction strength is that Q should be small and positive. If it had large and negative value then dark energy would have dominated the expansion practically from the outset and galaxies could not have formed at the desired epochs. Thus,

$$Q = \alpha\dot{\rho_m}. \tag{6}$$

where α is the proportionality constant. The interaction strength would, in general, depend on temperature (however, here we take it as independent of temperature). From (5) with form of Q as in (6), we have the following functional form of the energy densities of the components of tachyonic field

$$\rho_m = \rho_m^0\left(\frac{a}{a_0}\right)^{-3/(1-\alpha)}; \quad \rho_\lambda = \rho_\lambda^0 - \alpha\rho_m^0\left[\left(\frac{a}{a_0}\right)^{-3/1-\alpha} - 1\right] \tag{7}$$

where ρ_m^0 and ρ_λ^0 are the values of energy density of dark matter and dark energy respectively at present epoch scale factor a_0. It can be seen that in the absence of interaction ($\alpha = 0$) we have $\rho_m = \rho_m^0(a/a_0)^{-3}$ and $\rho_\lambda = \rho_\lambda^0$ = constant, as expected in the standard approach with a truly constant cosmological constant.

3. Constraints on interaction strength from observations

The difference of the squares of the normalized Hubble parameter $E^2(x)$ at two different redshifts x_i and x_j for the interaction term $Q = \alpha\dot{\rho_m}$ is given as

$$E^2(x_i) - E^2(x_j) = \triangle E^2(x_i, x_j) = (1-\alpha)\Omega_m^0\left[x_i^{3/1-\alpha} - x_j^{3/1-\alpha}\right]. \tag{8}$$

Measuring the values of $\triangle E^2(x_i, x_j)$ from the redshift observations and Ω_m^0 as the concordance value from CMBR, BAO etc. data analysis, we calculate α for the (x_i, x_j) pair of redshifts corresponding to interaction form Q. We emphasize that the choice of such pair is desirable over a large range of redshift observations, since we must determine the interaction strength over a long period of cosmic evolution. This is required so as to constrain the structure formation both at high- and low-z epochs. The precise measurements of H_0 will be helpful to break the degeneracy among the cosmological parameters [13]. Here, we take a set of values of $H(z)$ [14,15] as: at $z = 0.1$, 0.4, 1.3, and 1.75, the values of $H(z) = 69 \pm 12$, 95 ± 17, 168 ± 17 and 202 ± 40 km s^{-1} Mpc^{-1} respectively. Further calculations may be done by taking the present values $H_0 = 73.8 \pm 2.4 \approx 73.8$ km s^{-1} Mpc^{-1} [16] and the density parameter of matter component $\Omega_m^0 = 0.272$ [17]. Thus, choosing the four epochs as $x_1 = 1.1$, $x_2 = 1.4$, $x_3 = 2.3$ and $x_4 = 2.75$ for these redshifts, the squared normalized Hubble parameter may be calculated in a straightforward manner as given below:
$E^2(x_1) = 0.874$, $E^2(x_2) = 1.657$, $E^2(x_3) = 5.182$, and $E^2(x_4) = 7.492$. and thus $\Delta E^2(x_1, x_2) = -0.783$, $\Delta E^2(x_1, x_3) = -4.308$, $\Delta E^2(x_1, x_4) = -6.618$, $\Delta E^2(x_2, x_3) = -3.525$, $\Delta E^2(x_2, x_4) = -5.835$ and $\Delta E^2(x_3, x_4) = -2.310$.
Using these data we can proceed to determine six values of the proportionality constant α of the interaction form $Q = \alpha \dot{\rho_m}$ from (8). The solution of these six equations helps in estimation of α.

4. Entropy of the Universe

The entropy of the FRW universe includes two terms– one is the entropy of the apparent horizon and the second is entropy contributed by fluids enclosed by the horizon. First, let us consider entropy of the apparent horizon. The entropy of the event horizon is proportional to its surface area A [18] and is given as

$$S_A = \frac{k_B}{4} \frac{A}{l_{pl}^2}. \tag{9}$$

where $A = 4\pi r_A^2$ is the area of the apparent horizon and $r_A = \left(\sqrt{H^2 + \frac{k}{a^2}}\right)^{-1}$ is radius of the horizon. Also, l_{pl} and k_B are Planck's length and Boltzmann's constant respectively. Using the Friedmann equations $H = \frac{8\pi G}{3}[\rho_m + \rho_\lambda]$ (for $k = 0$ universe) we have the following expression for the apparent horizon

$$A = \frac{3}{2G}[\rho_m^0 x^{3/1-\alpha} + \rho_\lambda^0 - \alpha \rho_\lambda^0 (x^{3/1-\alpha} - 1)]. \tag{10}$$

Hence, the entropy of apparent horizon is

$$S_A = \frac{3k_B}{8Gl_{pl}^2}[\rho_m^0 x^{3/1-\alpha} + \rho_\lambda^0 - \alpha \rho_\lambda^0 (x^{3/1-\alpha} - 1)]. \tag{11}$$

The entropy of fluids enclosed by the apparent horizon can be calculated as follows. If there are N dust particles and each particle contributes an entropy equal to k_B, the total entropy of fluid inside the apparent horizon will be equal to $S_m = k_B.N$ with $N = \left(\frac{4\pi}{3} r_A^3\right) n$ where n is the number density of the dust particles. The conservation law of the number density gives $n = n_0 \left(\frac{a_0}{a}\right)^3 = n_0 x^3$. Thus, the entropy contributed by fluids turns out to be

$$S_m = k_B \frac{4\pi}{3} r_A^3 n_0 x^3 \tag{12}$$

whereas the total entropy of universe becomes equal to

$$S_A + S_m > 0. \tag{13}$$

The interaction between components inside the apparent horizon also produces entropy and must contribute to the observed large amount of entropy in the universe.

5. Conclusion

The cosmic tachyonic scalar field may be decomposed into several components with the assumption that the field is spatially homogeneous. Here we take two components, one is pressure-less dust matter and other is the cosmological constant (because its equation of state is -1). The non-equilibrium state of two components has a tendency to undergo a transition into the equilibrium state and to achieve this state, these components enter into mutual interaction. During interaction the conservation of energy of individual components gets violated while the total energy of field is held conserved. The thermodynamic laws responsible for the transfer of energy between components are obeyed with the heuristic choice of interaction strength Q. Normalized Hubble parameter at different redshifts provided by the observations helps to constrain the interaction strength Q with the calculation of α. The entropy of universe appears as a sum of entropy of apparent horizon and entropy of fluid inside apparent horizon. We assume the fluid inside the apparent horizon is dust having finite number of particles.

6. Acknowledgments

The authors thankfully acknowledge the support from the University Grants Commission, New Delhi for the present work through the Major Research Project vide F. No.37-431/2009 (SR).

References

[1] Riess A G et al. 1998 *Astron. J.* **116** 1009
[2] Perlmutter S et al. 1999 *Astrophys. J.* **517** 565
[3] Komatsu E et al. 2011 *Astrophys. J.S.* **192** 18
[4] Blake C and Glazebrook K 2003 *Astrophys. J.* **594** 665
[5] Sen A 2002 *JHEP* **0204** 048
[6] Padmanabhan T and Roy Choudhury T 2002 *Phys. Rev.* D **66** 081301
[7] Amendola L 2000 *Phys. Rev.* D **62** 043511
[8] Zimdahl W and Pavon D 2011 *Phys. Lett.* B **521** 133
[9] Verma M M 2010 *Astrophys. Space Sci.* **330** 101
[10] Chimento L P 2010 *Phys. Rev.* D **81** 043525
[11] Callen H B 1960 *Equilibrium Thermodynamics* (New York: Wiley)
[12] Keizer J 1987 *Statistical Thermodynamics of Non-equilibrium Processes* (New York: Springer)
[13] Freedman W L and Madore B F 2010 The Hubble Constant *Annu. Rev. Astron. Astrophys.* **48** *Preprint:* astro-ph/1004.1856
[14] Simon J, Verde L and Jimenez R 2005 *Phys. Rev.* D **71** 123001
[15] Stern D et al. 2010 *JCAP* **1002** 008
[16] Riess A G, Macri L, Casertano S, Lampeitl H, Ferguson H C, Filippenko A V, Jha S W, Li W and Chornock R 2011 *Astrophys. J.* **730** 119
[17] Larson D et al. 2011 *Astrophys. J.S.* **192** 16
[18] Radicella N and Pavon D 2012 *Gen Relativ Gravit* **44** 685

Neutrino masses and mixings using updated values of running quark and lepton masses

Kalpana Bora and Gayatri Ghosh
Department Of Physics, Gauhati University, Assam, India
E-mail: kalpana@gauhati.ac.in
E-mail: gayatrighsh@gmail.com

Abstract. Running values of quark and lepton masses at higher scales in Grand Unified Theories are very important, for building models of fermion masses, and to calculate neutrino masses and mixings. In this work, we will calculate neutrino masses and mixings at GUT scales, using updated values of running quark and lepton masses in SO(10) Grand Unified Theory.

1. Introduction

Current research in experimental and theoretical High Energy Physics aims at explaining the origin of all fermions masses and mixings including those of neutrinos. Neutrinos have masses and mix with each other and oscillate to other flavours. Grand Unified theories (GUTs) like SO(10) which unify strong and electroweak interactions are considered as prime candidates for describing neutrino masses [1,2]. In SO(10) all the known fermions of a given generation including the right handed neutrino are present in a single sixteen dimensional spinorial representation of the group. These theories require running masses and mixings of quarks and charged leptons at GUT scales for calculating neutrino masses. SO(10) GUT has a number of desirable features which helps in describing physics beyond the Standard Model. Non zero neutrino masses, the see-saw scale and a high B-L scale fits naturally in a grand unified model group based on the gauge group SO(10). SO(10) model contains the L-R symmetric unification group $SU(2)_L \times SU(2)_R \times U(1)_{B-L} \times SU(3)_C$ which provides the origin of parity violation in nature. In a class of minimal SO(10) model both the Dirac neutrino masses and Majorana neutrino masses are related to observables in the charged fermion sector. We use a simple Higgs system with one 10 and 126 which have Yukawa couplings to Fermions $16_F(Y_{10}10_H + Y_{126}126_H)$ [3]. The 10 is for quark and lepton masses, 126 is required for seesaw mechanism. The breaking of SO(10) via G_{224} is achieved by either 54 or 210 dimensional Higgs. The Fermion Higgs coupling of the SO(10) model can be written down as in Ref.[2]

$$L_Y = h_{ab}\psi_a\psi_b H + f_{ab}\psi_a\psi_b\bar{\Delta} + H.C \tag{1}$$

Where H is the 10 dimensional Higgs of SO(10) .$\bar{\Delta}$ is the 126-plet of Higgs. Here ψ_a , a=1-3 denote the three families of fermions belonging to the 16 dimensional representation of SO(10). Under G_{224},

$$\overline{126} = (1, 1, 6) + (1, 3, 10) + (3, 1, \overline{10}) + (2, 2, 15) \tag{2}$$

$$10 = (1,1,6) + (2,2,1) \tag{3}$$

where (1,3,10) and (2,2,15) component is denoted by Δ_R and Σ respectively. $<\Delta_R^o>= v_R = 10^{12}$ GeV breaks the intermediate symmetry down to the SM and generates Majorana neutrino mass. The scalar potential contains a crucial term [2]

$$V = \lambda\Delta\overline{\Delta}\Delta H \tag{4}$$

which induces vevs for the standard doublets in the Σ multiplet of 126. In above SO(10) model , the neutrino mass matrix is given by the type II seesaw formula, that contains the induced triplet vev which dominates the neutrino mass matrix. Under certain assumptions, we have [4]

$$M_\nu = a(m_l - m_d) \tag{5}$$

The paper is organized as follows. In section II we give formula for type II seesaw mechanism using induced vev. In section III we calculate neutrino oscillation parameters using equation (5). In sec IV we give our conclusions.

2. TYPE II SEESAW MECHANISM

The $SO(10)$ invariant Superpotential giving the Yukawa Couplings of the 16 dimensional matter spinor ψ_i with the Higgs set $H_{10} = 10$ and $\Delta = \overline{126}$ is [4]

$$W_Y = h_{ij}\psi_i\psi_j H_{10} + f_{ij}\psi_i\psi_j\overline{\Delta} \tag{6}$$

where h and f are symmetric matrices . In the type I seesaw mechanism the neutrino mass matrix is given by

$$M_\nu = -M_\nu^D M_{NR}^{-1} M_\nu^{TD} \tag{7}$$

where $M_{NR} = f v_{B-L}$ and v_{B-L} is the see saw scale . In SO(10) the true seesaw formula has a second term which comes from an induced $SU(2)_L$ triplet vev . The type II see saw mechanism has the following form [4]

$$M_\nu = f\sigma_L - M_\nu^D M_{NR}^{-1} M_\nu^{TD} \tag{8}$$

$$\sigma_L = \lambda\frac{v^2}{v_{B-L}} \tag{9}$$

v is the $SU(2)_R$ breaking scale. λ is a combination of parameters in Higgs potential. We note the decomposition of $\overline{126}$ under the group $SU(2)_L \times SU(2)_R \times SU(4)_C$

$$\overline{126} = (1,1,6) + (2,2,15) + (3,1,\overline{10}) + (1,3,10) \tag{10}$$

The $SU(2)_L$ triplet $\Delta_L = (3,1,\overline{10})$ couples to the left handed multiplet $\psi_L = (2,1,4)$ of the 16 dimensional SO(10) containing the matter spinor i.e $\psi_L\psi_L\Delta_L$. The mass of the right handed neutrino originates from the coupling of $\Delta_R = (1,3,10)$ of $\overline{126}$ to right handed fermion multiplet $\psi_R^C = (1,2,4)$ i.e $\psi_R\psi_R^C\Delta_R$ [5]. Δ_R produces the second term in the type II seesaw formula. The sumrule (eq 5) holds when triplet vev contribution to the neutrino mass matrix dominates in the type II seesaw mechanism.

3. CALCULATION OF NEUTRINO OSCILLATION PARAMETERS

In our calculation of neutrino oscillation parameters,we use the updated values of running quark and lepton masses from Ref.[6] and are given in table 1. Then using equation (3) we calculate neutrino oscillation parameters, which are presented in figure 1.

$Fermion$	$Mass$
m_u	$0.3963^{+0.1506}_{-0.1284} MeV$
m_c	$0.1932^{+0.0240}_{-0.0246} Gev$
m_t	$80.4472^{+2.9128}_{-2.6158} GeV$
m_d	$0.9284^{+0.3838}_{-0.3754} MeV$
m_s	$17.6097^{+4.8972}_{-4.6737} MeV$
m_b	$1.2424^{+0.0626}_{-0.0572} GeV$
m_e	$0.3569^{-0.0001}_{+0.0001} MeV$
m_μ	$75.3570^{-0.0744}_{+0.0682} MeV$
m_τ	$1.6459^{+0.0114}_{-0.0206} GeV$

Table 1: Values of Masses in MSSM, 2 Loop,tan β =55 [UPDATED] Fermion masses used from [6] at GUT scale = $2 \times 10^{16} GeV$

In figure 1 in the first figure we have shown our results for $sin^2\theta_{23}$ and Δm^2_{23}. In the second figure we have shown the results for $sin^2\theta_{12}$ and Δm^2_{12} and in the third figure the results are given for Δm^2_{23} and θ_{13}. We know the latest values of global fit values of neutrino oscillation parameters are [7]

$$\Delta m^2_{21} = 7.54 \times 10^{-5} eV^2$$

$$\Delta m^2_{31} = 2.43 \times 10^{-3} eV^2$$

$$sin^2\theta_{12} = 0.307$$

$$sin^2\theta_{13} = 0.0241$$

$$sin^2\theta_{23} = 0.427$$

We find that values of oscillation parameters as calculated by us (in fig 1) agree well with latest global fit values.

4. CONCLUSION

To conclude, in this work we have calculated neutrino oscillation parameters in minimal SO(10) theory, using updated values of running quark and lepton masses. We find that these values agree with latest global fit values as given in Ref. [7] of these parameters. We would like to do R.G.E study of the neutrino oscillation parameter in our future work.

Journal of Physics: Conference Series **481** (2014) 012016 doi:10.1088/1742-6596/481/1/012016

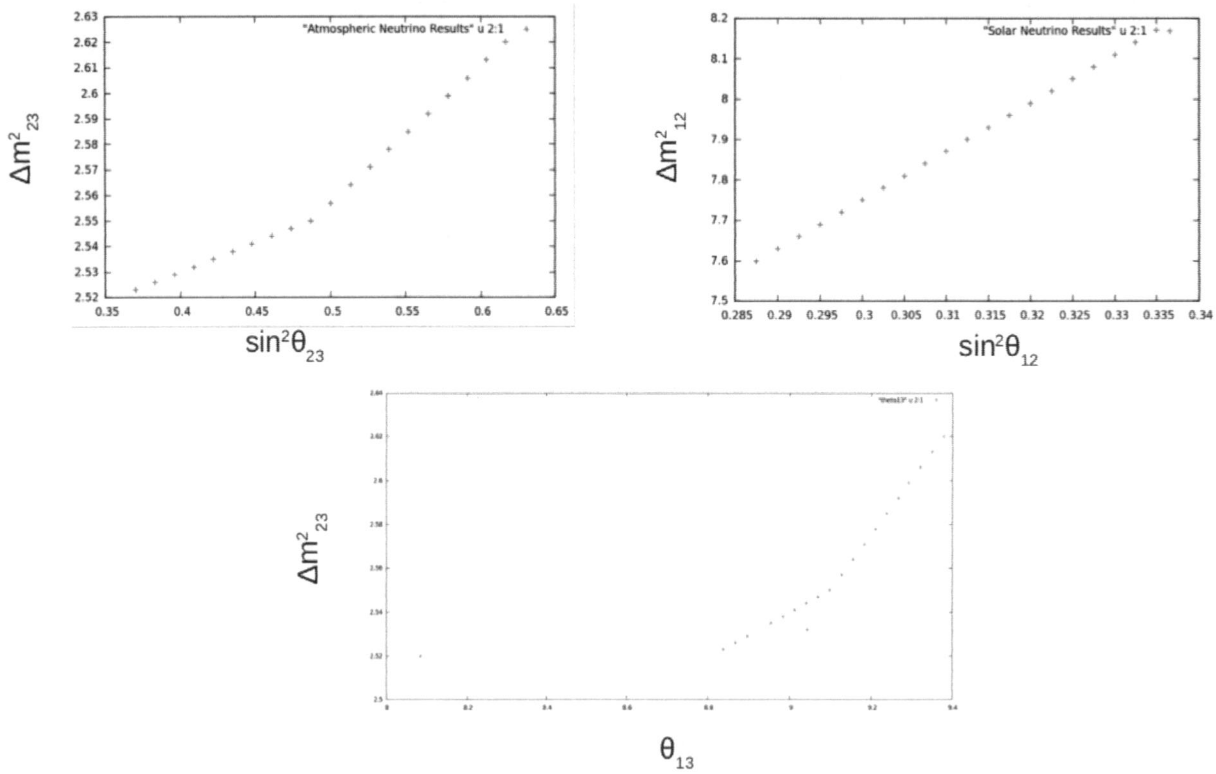

Figure 1: Curves are plotted within 1 σ errors of the best fit values of Δm^2_{23} and $sin^2\theta_{23}$, Δm^2_{12} and $sin^2\theta_{12}$,Δm^2_{23} and θ_{13} respectively.

References

[1] K.S.Babu, R.N. Mohapatra, arXiv:hep-ph/9209215v1,Phys.Rev.Lett.70(1993) 2845-2848
[2] Dae-Gyu Lee,R.N.Mohapatra, arxiv:hep-ph/9406328v3,Phys.Rev.D51(1995) 1353-1361
[3] Anjan S. Joshipura and Ketan M Patel arXiv:1102.5148v1 [hep-ph] 25 Feb 2011 Phys.Rev.D83:095002
[4] H. S. Goh, R. N. Mohapatra, S-P g, hep-ph/0303055
[5] H. S. Goh, R. N. Mohapatra, S-P Ng, Phys. Rev D68, 115008 (2003).
[6] Kalpana Bora, arXiv:1206.5909v1[hep-ph] 26 Jun2012 (Horizon), Vol 2 ,ISSN No 2250-0871 ,2013
[7] G.Fogli, E. Lisi, A. Marrone, D. Montanino, A. Palazzo, et. al., Phys.Rev.D86(2012) 013012, [arXiv:1205.5254]

Discovery reach of CP violation in neutrino oscillation experiments with standard and non-standard interactions

Zini Rahman[1,*], Arnab Dasgupta[1,†], and Rathin Adhikari[1,††]
[1]Centre for Theoretical Physics, Jamia Millia Islamia, Jamia Nagar, New Delhi 110025, India.
E-mail: *zini@ctp-jamia.res.in †arnab@ctp-jamia.res.in ††rathin@ctp-jamia.res.in

Abstract. Considering the value of mixing angle θ_{13} as obtained in recent Daya Bay and MINOS experiments we have explored the discovery reach of CP violation in the leptonic sector due to complex δ phase in neutrino oscillation experiments with superbeam. We have considered two experimental set-up corresponding to two baselines of 130 Km and 2300 Km respectively. Considering numerical simulation we have shown how the discovery reach of CP violation will change due to the presence of various non-standard interactions (NSIs). Although it is known that short baseline is better choice for CP violation study for Standard Model (SM) interactions of neutrinos with matter but in presence of NSI we find that sometimes better discovery reach is possible in relatively longer baselines.

1. Introduction

Among the various neutrino oscillation parameters the values of two mixing angles θ_{12} and θ_{23} have been provided by experiments with certain accuracy. Regarding the third mixing angle θ_{13}, recently the Daya Bay experiment [1] and MINOS [2] has found non-zero large value of $\sin^2 2\theta_{13}$. The magnitude of the mass squared differences $|\Delta m^2_{31}|$ and Δm^2_{21} are also known but the sign of Δm^2_{31}(hierarchy) and the CP violating phase δ are still unknown. In this work we try to explore CP violation due to δ in the light of recent experimental findings of large value of θ_{13} and non-standard interactions(NSIs) [3, 4]. For this we have considered a superbeam source and two different baselines one of length 2300 Km and another of 130 Km length.

2. NSI and its effect in neutrino oscillation probabilities for different baselines

In addition to the Standard Model (SM) Lagrangian density we consider the following non-standard fermion-neutrino interaction in matter defined by the Lagrangian:

$$\mathcal{L}^M_{NSI} = -2\sqrt{2}G_F\varepsilon^{fP}_{\alpha\beta}[\bar{f}\gamma_\mu Pf][\bar{\nu}_\beta\gamma^\mu L\nu_\alpha] \tag{1}$$

where $P \in (L, R)$, $L = (1-\gamma^5)/2$, $R = (1+\gamma^5)/2$, $f = e, u, d$ and $\varepsilon^{fP}_{\alpha\beta}$ denotes the non-standard interactions(NSIs) [5, 6]. These NSI parameters can be reduced to the effective parameters as $\varepsilon_{\alpha\beta} = \sum_{f,P}\varepsilon^{fP}_{\alpha\beta}\frac{n_f}{n_e}$, where n_f is the fermion number density and n_e is the electron number density. These NSIs can modify the interaction of neutrinos with matter and thus change the oscillation probability of different flavor of neutrinos. In section 4 in presenting results of our

Journal of Physics: Conference Series **481** (2014) 012017 doi:10.1088/1742-6596/481/1/012017

numerical analysis considering NSI effect during propagation of neutrinos only, we consider the constraints on the NSIs as $|\varepsilon_{ee}| \leq 0.75$ [7], $|\varepsilon_{e\mu}| \leq 3.8 \times 10^{-4}$ [7], $|\varepsilon_{e\tau}| \leq 0.25$ [7], $|\varepsilon_{\mu\mu}| \leq 3 \times 10^{-2}$ [8], $|\varepsilon_{\mu\tau}| \leq 6 \times 10^{-3}$ [8] and $|\varepsilon_{\tau\tau}| \leq 3 \times 10^{-2}$ [8].

2.1. Oscillation Probability

For understanding qualitatively the NSI effect in finding the discovery reach of CP violation we consider the oscillation probability $P_{\nu_e \to \nu_\mu}$ given in equations (8.3) and (8.4) of ref [9] as $\nu_\mu \to \nu_e$ and $\bar{\nu}_\mu \to \bar{\nu}_e$ oscillation channels are particularly sensitive to CP violation. We can obtain $P_{\nu_\mu \to \nu_e}$ from $P_{\nu_e \to \nu_\mu}$ by replacing δ by $-\delta$.

To estimate the order of magnitude of different terms in the above mentioned oscillation probabilities we shall consider the average earth matter density of 0.0263 gm/cc and 3.314 gm/cc for a short baseline of 130 Km and and a relatively longer baseline of 2300 Km respectively and as such $A < \alpha$ for 130 Km baseline for $E \sim 0.3$ GeV and $A \sim 0.5$ for 2300 Km baseline for $E \sim 5$ GeV near the peak of the oscillation probability. Here, $A = 2E\sqrt{2}G_F n_e/(\Delta m^2_{31})$ and $\alpha = \Delta m^2_{21}/(\Delta m^2_{31})$ where $\Delta m^2_{ij} = m^2_i - m^2_j$, m_i denotes the mass of the i-th neutrino, G_F is the Fermi constant and n_e is the electron number density of matter. A is due to the interaction of neutrinos with matter in SM. For only SM interactions, (i.e $\varepsilon_{\alpha\beta} \to 0$) in equations(8.3)and (8.4) of ref [9], one finds that for both the baselines the δ dependence occurs at order of $\alpha^{3/2}$. However, for large values of A for longer baseline the matter effect is more in the δ independent part. For this reason the discovery reach of CP violation is better in short baseline.

However, when NSIs are also taken into account one can see that for longer baseline further δ dependence in $P_{\nu_\mu \to \nu_e}$ could occur at the order of $\alpha^{3/2}$ through terms in equations (8.3) and (8.4) containing NSIs - $\varepsilon_{\epsilon\mu}$ and $\varepsilon_{e\tau}$ of the order of α. We have checked that for slightly higher NSIs of the order of $\sqrt{\alpha}$ the same δ dependent terms in $P_{\nu_\mu \to \nu_e}$ appears with NSIs - $\varepsilon_{\epsilon\mu}$ and $\varepsilon_{e\tau}$ in the oscillation probability for long baseline making these terms at the order of α which could compete with the δ independent but matter dependent part in $P_{\nu_\mu \to \nu_e}$ as that is also at the order of α. This improvement of δ dependent part over independent part for long baseline does not happen for short baseline as δ dependent terms containing NSIs are small for short baseline due to small value of A. So presence of NSIs - $\varepsilon_{e\mu}$ and $\varepsilon_{e\tau}$ in δ dependent terms improves the discovery reach of CP violation for longer baseline. It is expected that in presence of these NSIs i.e. $\varepsilon_{e\mu}$ and $\varepsilon_{e\tau}$ the long baseline could provide a better discovery reach for CP violation. However, as the upper bound of $\varepsilon_{e\mu}$ is somewhat smaller one may expect $\varepsilon_{e\tau}$ will have significant effect in the discovery reach of CP violation. In fact, our numerical analysis also shows this feature.

3. Experimental setup, systematic uncertainties and errors of various parameters

In this work for the numerical simulation we consider two set-ups: (a) A superbeam setup which originates in CERN and reaches a 100 kt Liquid Argon detector placed at a distance of 2300 Km at Pyhäsalmi (Finland) (b) A Superbeam setup originating in CERN and reaching a 500 Kt Water Cherenkov detector [10] placed at a distance of 130 Km at Fréjus (France).

For both the set-ups the true values of the neutrino oscillation and the errors on these parameters are considered as in ref [11]. In doing the whole analysis we have used GLoBES software [12, 13].

For set-up (a), in doing the numerical analysis to study the response of the detector we have followed ref [11]. The time period has been taken to be 5 years each for neutrinos and anti-neutrinos. The correlation between the visible energy of background NC events and the neutrino energy is implemented by migration matrices which has been provided by L. Whitehead [14].

The flux considered for set-up (b) has mean energy ~ 0.3 GeV, for 3.5 GeV protons and 10^{23} protons on target per year. The beam power has been considered of about 4 MW per year and the time period has been taken to be 2 years for neutrinos and 4 years for anti-neutrinos.

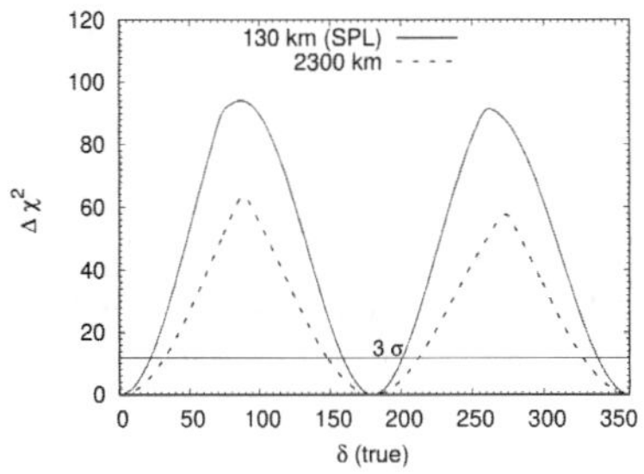

Figure 1: Discovery reach of CP violation due to δ for two different baselines 130 Km and 2300 Km considering only SM interactions and NH at 3σ.

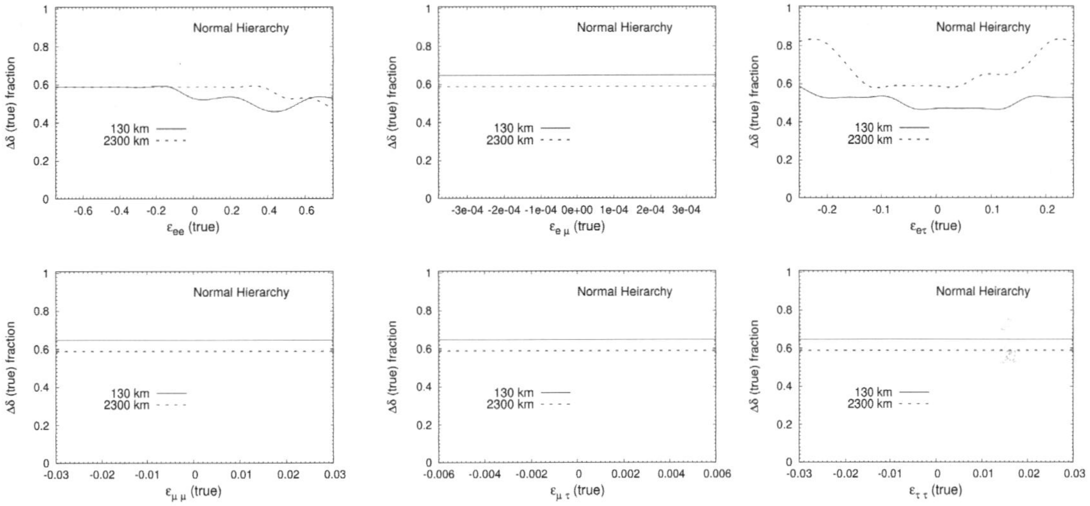

Figure 2: Fraction of δ for CP violation discovery ($\Delta\delta$ fraction) for two different baselines 130 Km and 2300 Km at 3σ considering NSIs for NH.

We consider the same flux as in [15, 16]. The efficiencies for the signal and background are included in the migration matrices based on [10] except for the channels ν_μ disappearance, $\bar{\nu}_\mu$ disappearance and ν_μ (NC) which are 64%, 81% and 11.7% efficiencies respectively. We have considered systematic uncertainties of 2% on signal and background channels.

4. Discovery reach of CP violation due to δ for real NSIs'

In this work we have done a comparative study for the two experimental set-ups (a) and (b) in finding the discovery reach of the CP violation due to phase δ depending on different NSI parameters (true values) and for true hierarchies.

In figure 1 we have shown the the discovery reach of CP violation for SM interactions of neutrinos with matter. Particularly at 3σ confidence level for normal hierarchy(NH) the CP violation could be discovered over about 78% of the possible δ values for 130 Km and 65% for 2300 Km baselines. The discovery reach for 2300 Km was shown earlier by Coloma *et al* [11]. So with only SM the short baseline like 130 Km seems to give better discovery reach of CP violation. This was observed earlier by different authors [17–21]. However, the short baseline may not be always better in presence of NSI which we discuss below.

In figure 2 we have compared the discovery reach of CP violation for 130 Km baseline and 2300 Km baseline in presence of NSIs at 3σ confidence level for NH considering only one NSI at a time. Here we define $\Delta\delta$ fraction as the ratio of sum of all ranges of δ (true) values for which

CP violation could be discovered divided by total allowed range of δ (true) between 0 to 2π. This fraction corresponds to discovery reach of CP violation.

For ε_{ee} in the range about -0.2 to 0.6 the discovery reach of CP violation due to δ is found to be better for 2300 Km baseline.. However for $\varepsilon_{e\tau}$ in its almost entire allowed range for 2300 Km baseline there is better discovery reach. In presence of NSIs — $\varepsilon_{e\mu}$, $\varepsilon_{\mu\mu}$, $\varepsilon_{\mu\tau}$ and $\varepsilon_{\tau\tau}$ we see for the entire allowed range of these NSI parameters there is better discovery reach for 130 Km baseline. We have checked the discovery of CP violation for inverted hierarchy(IH) also and we find similar results with NH.

One may note that the effect of the NSI — $\varepsilon_{e\mu}$, $\varepsilon_{\mu\mu}$, $\varepsilon_{\mu\tau}$ and $\varepsilon_{\tau\tau}$ on the CP violation discovery is negligible due to their relatively smaller values. However, statistically due to the inclusion of extra NSI parameter the degree of freedom has increased thus making the discovery reach fraction little bit lesser in comparison to the case with only SM interaction. The straight line in the figures 2 corresponding to those NSIs' indicates this negligible effect.

5. Conclusion

Considering the large value of θ_{13} provided by the Daya Bay experiment we have tried to study the CP violation discovery reach using a short baseline of length 130 Km and a relatively longer baseline of 2300 Km. It is found that in the presence of only SM interaction shorter baseline is better and if NSIs' are considered longer baselines also could give better discovery reach of CP violation. Considering the possibility of the presence of NSIs in nature it seems both short and long baseline should be considered particularly for discovery of CP violation in the leptonic sector through neutrino oscillation experiment.

6. Acknowledgment

AD thanks Council of Scientific and Industrial Research, India for financial support through Senior Research Fellowship and ZR thanks University Grants Commission, Govt. of India for providing research fellowships.

References

[1] F. P. An *et al.* [DAYA-BAY Collaboration] 2012 *Phys. Rev. Lett.* **108** 171803.
[2] P. Adamson *et al.* [MINOS Collaboration] *Phys. Rev. Lett.* . (*Preprint* arXiv:1301.4581).
[3] J. Kopp, M. Lindner, T. Ota and J. Sato 2008 *Phys. Rev. D* **77** 013007.
[4] T. Ota 2008 *AIP Conf. Proc.* V 981 (United States) p231.
[5] A. M. Gago, H. Minakata, H. Nunokawa, S. Uchinami and R. Zukanovich Funchal 2010 *J. High Energy Phys.* JHEP **1001** 049.
[6] G. Mitsuka *et al.* [Super-Kamiokande Collaboration] 2011 *Phys. Rev. D* **84** 113008.
[7] T. Ohlsson 2011 *Rept. Prog. Phys.* **76** 044201.
[8] A. Esmaili and A. Y. .Smirnov 2013 *J. High Energy Phys.* JHEP **1306** 026.
[9] K. Asano and H. Minakata 2011 *J. High Energy Phys.* JHEP **1106** 022.
[10] L. Agostino *et al.* [MEMPHYS Collaboration] 2013 *J. Cosmol. Astropart. Phys.* JCAP **1301** 024.
[11] P. Coloma, E. Fernandez-Martinez and L. Labarga 2012 *J. High Energy Phys.* JHEP **1211** 069.
[12] P. Huber, M. Lindner and W. Winter 2005 *Comput. Phys. Commun.* **167** 195.
[13] P. Huber, J. Kopp, M. Lindner, M. Rolinec and W. Winter 2007 *Comput. Phys. Commun.* **177** 432.
[14] L. Whitehead (Private communication).
[15] J. -E. Campagne, M. Maltoni, M. Mezzetto and T. Schwetz 2007 *J.High Energy Phys.* **0704**, 003.
[16] http://www.mpi-hd.mpg.de/personalhomes/globes/glb/SPL.html
[17] S. K. Agarwalla, T. Li and A. Rubbia 2012 *J. High Energy Phys.* JHEP **1205** 154.
[18] P. Coloma, A. Donini, E. Fernandez-Martinez and P. Hernandez 2012 *J. High Energy Phys.* JHEP **1206** 073.
[19] P. Coloma, E. Fernandez-Martinez and L. Labarga 2012 *J. High Energy Phys.* JHEP **1211** 069.
[20] P. Coloma, T. Li and S. Pascoli *Preprint* arXiv:1206.4038.
[21] P. Coloma, P. Huber, J. Kopp and W. Winter *Preprint* arXiv:1209.5973.

Probing the type-III seesaw model through $e^+e^- \rightarrow \Sigma^+\Sigma^-$ at ILC

Sumit K Garg

Department of Physics and IPAP, Yonsei University, Seoul 120-749, Korea.

E-mail: sumit@yonsei.ac.kr

Deepanjali Goswami

Department of Physics, Indian Institute of Technology Guwahati, Guwahati 781039, Assam, India.

E-mail: deepanjali.goswami@iitg.ernet.in

P Poulose

Department of Physics, Indian Institute of Technology Guwahati, Guwahati 781039, Assam, India.

E-mail: poulose@iitg.ernet.in

Abstract. We investigate the type-III seesaw fermionic charged triplet ($\Sigma^\pm$) production in the process $e^+e^- \rightarrow \Sigma^+\Sigma^-$ at International Linear Collider (ILC), with further decay of Σ to $W\nu$ and to $H\mu$. The final state decay distributions of triplets are analysed in order to check the sensitivity of their mixing with SM leptons and to understand their properties. The triplet state (Σ) having mass around 200 GeV, will be produced in large numbers at ILC with centre of mass energy of 500 GeV and a moderate integrated luminosity of 50 fb $^{-1}$.

1. Introduction

Neutrino oscillation experiments have confirmed that neutrinos have mass, albeit many orders of magnitude smaller than the charged lepton masses. This opens a new window beyond the Standard Model (SM), proposing various different ways to accommodate this observation.The seesaw mechanism is one of the most attractive methods in which a small Majorana mass can be generated for each of the known neutrinos in the presence of new heavy states. It explains the smallness of neutrino mass with the help of a suppression factor $\frac{v}{M}$, which is the ratio of electroweak scale to a new scale. The seesaw models are of three different types, called type-I [1], type-II [2] and type-III [3] introducing heavy states of mass M, that involves weak isospin singlets, scalar triplets and fermion triplets,respectively. The new particles predicted by the seesaw mechanism can couple to the SM particles to produce virtual contribution to processes involving ordinary particles in the initial and final states, which may lead to observable effects beyond the light neutrino sector. In this work we will focus on the type-III seesaw model. Some recent work on the phenomenology of type-III seesaw in the context of Large Hadron Collider(LHC) may be found in Ref. [4] and references therein. The suppression factor of

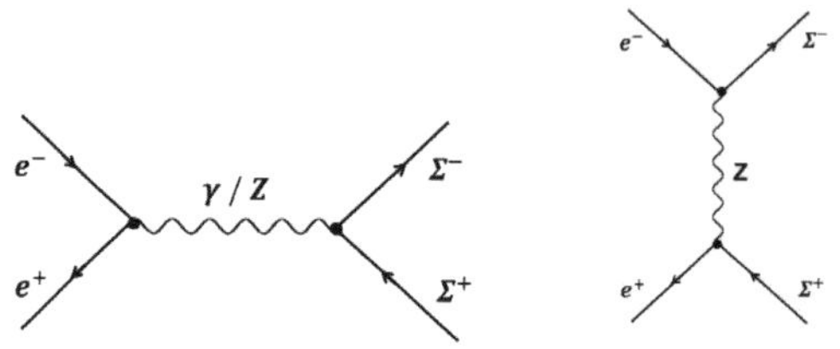

Figure 1. Feynman diagrams contributing to the process $e^+e^- \to \Sigma^+\Sigma^-$ in type-III seesaw model.

v/M,where v is the vacuum expectation value of the Higgs field and M is the mass of the heavy particle, implies $M \sim 10^{14}$GeV so as to generate small neutrino masses($m_\nu \sim meV$) with Yukuwa couplings of order one. However in principle this scale can be as low as $M \sim 100$ GeV which can be easily tested at current and future colliders. With the upcoming International Linear Collider (ILC) expected to run at a few hundred GeV to a TeV or more centre of mass energy, it is possible to produce the heavy new particles predicted by this model. For phenomenological studies in the context of ILC, please see Ref. [5]. In this Letter we will study the production of the heavy charged leptons at ILC with their subsequent decays.

2. The model

The type-III seesaw model has, in addition to the SM fields, SU(2) triplets of fermions with zero hypercharge, Σ. While one such triplet is enough to generate mass for one of the neutrinos, more triplets are necessary to have non vanishing masses for at least two neutrinos, as is clear from neutrino oscillation studies. Although it is true that in the realistic case we have more than one triplets of fermions, in the present discussion we consider the presence of one triplet in the physical spectrum in the GeV and TeV scale. The other triplets may be imagined to be very heavy, and therefore decoupled from the physical spectrum at these energy scales. In terms of the compact 2×2 notation for triplets the Lagrangian involving these fields is given by [6]

$$\mathcal{L} = \mathrm{Tr}[\bar{\Sigma} \not{D} \Sigma] - \tfrac{1}{2} Tr\left[\bar{\Sigma} M_\Sigma \Sigma^c + \bar{\Sigma}^c M_\Sigma^* \Sigma\right] - \tilde{\phi}^\dagger \bar{\Sigma} \sqrt{2} Y_\Sigma L - \bar{L} \sqrt{2} Y_\Sigma^\dagger \Sigma \tilde{\phi}, \qquad (1)$$

where M_Σ is the mass matrix of the triplet and Y_Σ is the Yukuwa coupling matrix, $L \equiv (l, v)^T$ is the SM left-handed doublet lepton field,$\phi \equiv (\phi^+, \phi^0)^T \equiv (\phi^+, (v+H+i\eta)/\sqrt{2}\ )^T$ is the Higgs field, $\tilde{\phi} = i\tau_2 \phi^*$. The fermion triplet and Σ and its conjugate $\Sigma^c \equiv C\bar{\Sigma}^T$are explicitly given by

$$\Sigma = \begin{pmatrix} \Sigma^0/\sqrt{2} & \Sigma^+ \\ \Sigma^- & -\Sigma^0/\sqrt{2} \end{pmatrix}, \quad \Sigma^c = \begin{pmatrix} \Sigma^{0c}/\sqrt{2} & \Sigma^{-c} \\ \Sigma^{+c} & -\Sigma^{0c}/\sqrt{2} \end{pmatrix}$$

The covariant derivative is expressed in terms of the gauge fields as

$$D_\mu = \partial_\mu - i\sqrt{2}\, g \begin{pmatrix} W_\mu^3/\sqrt{2} & W_\mu^+ \\ W_\mu^- & -W_\mu^3/\sqrt{2} \end{pmatrix}.$$

3. Results

For the convenience of presenting the experimental bounds, the parameters in the above Lagrangian are expressed in terms of new set of parameters, which essentially parametrizes the mixing between the new triplet leptons and the SM leptons as $\frac{v^2}{2} | Y^\dagger M^{-2} Y|_{\alpha\beta} = | V_\alpha V_\beta|$.

In Fig.1, we present the Feynman diagrams involving the production of Σ pair at ILC in the limit of vanishing electron mass. The s-channel mediated by the photon and Z Boson depends

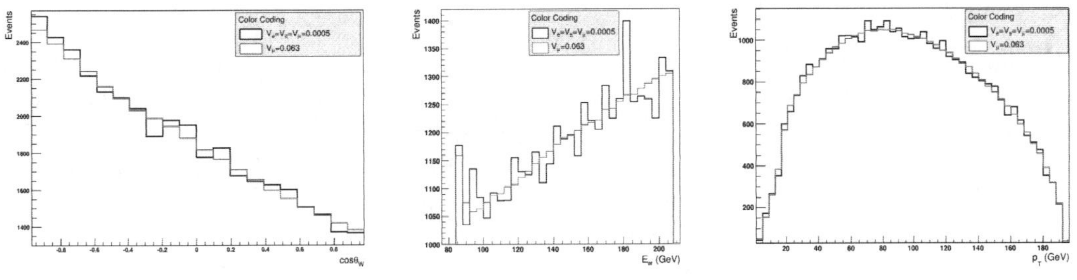

Figure 2. Angular, energy and p_T distributions of W^+ in the case of $W^+W^-\nu\nu$ final state, with c.m.energy, $\sqrt{s}$ =500 GeV and an integrated luminosity of 50 fb^{-1}.

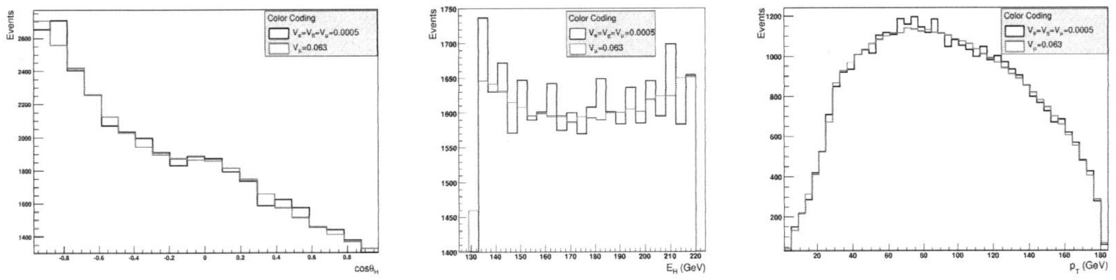

Figure 3. Angular, energy and p_T distributions of H in the case of $HH\mu^+\mu^-$ final state, with c.m.energy $\sqrt{s}$ =500 GeV and an integrated luminosity of 50 fb^{-1}.

on all the three mixing parameters, while the t-channel does depend on the value of V_e . Thus, analysis involving the angular distribution of the Σ produced is expected to enable an effective probe of this mixing parameter. We may note that this is very unlike similar studies at the Large Hadron Collider (LHC), where the production is completely insensitive to the mixing parameter. We consider further decay of the triplet fermion through the channels, $\Sigma^{\pm}\to W^{\pm}\nu$ and $\Sigma^{\pm}\to\mu^{\pm}H$, so that the complete process is given by

$$e^+e^- \to \Sigma^+\Sigma^- \;\to\; W^+W^-\nu\nu, \tag{2}$$
$$e^+e^- \to \Sigma^+\Sigma^- \;\to\; HH\mu^+\mu^-. \tag{3}$$

Coming to the present experimental limits on the parameters, if only one of V_e or V_μ or V_τ is present, the mixing parameters can be of the order of 10^{-2}, whereas with stronger constraints from flavour sector limits to the product of two couplings of the order of $10^{-4} - 10^{-7}$ when more than one of the mixing parameters are non-zero, as explained in Biggio and Bonnet in Ref. [6]. Present limits on the mass parameter M allows this to be larger than about 200 GeV. The process under scrutiny depends on this mass parameter in an obvious way. In the case of the mixing parameters, as mentioned above, the production process is expected to be sensitive to the parameter V_e . In the decay, on the other hand, it is a combination, $b_\alpha = \frac{|V_\alpha|^2}{|V_e|^2+|V_\mu|^2+|V_\tau|^2}$, described in the first reference in Ref. [7] which is relevant. Thus, if only one mixing is present, the decay is totally insensitive to this parameter. On the other hand, if more than one mixing is present, the decay could also be sensitive to these parameters,despite the fact that it is suppressed by two or three orders of magnitude smaller. Our main objective in this study is to investigate possibilities of establishing the new resonance, and further probing the mixings. In this report we present the results of the study so far. A more detailed analysis will be presented in a future publication. For illustrative purpose, we consider $M = 200$ GeV, and two sets of mixing values, viz., (i) $V_e = V_\tau = 0, V_\mu = 0.063$ and (ii) $V_e = V_\tau = V_\mu = 0.0005$.

Our numerical results are presented for the above parameters at an ILC of centre of mass energy of 500 GeV and for an integrated luminosity of 50 fb^{-1}. We first consider the process with Σ decaying into W Bosons. In Fig.2 we present the angular distribution for the final state W. Backgrounds are important, as complete reconstruction is not possible owing to the missing energy through neutrinos. Analysis of the background and the signal significance will be presented in our future work. Fig.2 also presents the energy and p_T distributions of the W. Coming to the other decay channel considered, we present in Fig.3 the angular distribution,the energy and p_T distribution of H Boson. Note that, in this case full event reconstruction is possible. We expect also the beam polarisation to be significant in the probe, as is clear from the t-channel process with a vertex involving both the initial electron and the heavy fermion. A detailed study involving all these aspects of this process is deferred to the future publication.

4. Conclusions

To summarise, we study the direct production and further decay of the heavy triplet fermions, Σ arising in a simplified version of the type-III seesaw model, which is one of the promising models with a realisable mechanism to generate small neutrino mass. In this preliminary report we have considered the cross section of the process and distribution of some of the final states for illustrative values of the parameters, and for typical ILC parameters. Detailed study involving beam polarisation and to understand the background are being carried out. Further, it is noted that the process has potential to probe the significance of different mixing parameters.

5. Acknowledgments

Work of DG and PP is partly supported by a BRNS, DAE, Government of India (Project No.: 2010/37P/49/BRNS/1446).

References

[1] P. Minkowski, Phys. Lett. B **67**, 421 (1977); R. N. Mohapatra and G. Senjanovic, Phys. Rev. Lett. **44**, 912 (1980).

[2] M. Magg and C. Wetterich, Phys. Lett. B **94**, 61 (1980); T. P. Cheng and L. -F. Li, Phys. Rev. D **22**, 2860 (1980); G. B. Gelmini and M. Roncadelli, Phys. Lett. B **99**, 411 (1981); G. Lazarides, Q. Shafi and C. Wetterich, Nucl. Phys. B **181**, 287 (1981); R. N. Mohapatra and G. Senjanovic, Phys. Rev. D **23**, 165 (1981).

[3] R. Foot, H. Lew, X. G. He and G. C. Joshi, Z. Phys. C **44**, 441 (1989); E. Ma, Phys. Rev. Lett. **81**, 1171 (1998) [hep-ph/9805219]; E. Ma and D. P. Roy, Nucl. Phys. B **644**, 290 (2002) [hep-ph/0206150]; T. Hambye, Y. Lin, A. Notari, M. Papucci and A. Strumia, Nucl. Phys. B **695**, 169 (2004) [hep-ph/0312203].

[4] R. Franceschini, T. Hambye and A. Strumia, Phys. Rev. D **78**, 033002 (2008) [arXiv:0805.1613 [hep-ph]]; F. del Aguila and J. A. Aguilar-Saavedra, Nucl. Phys. B **813**, 22 (2009) [arXiv:0808.2468 [hep-ph]]; B. He, N. Okada and Q. Shafi, Phys. Lett. B **716**, 197 (2012) [arXiv:1205.4038 [hep-ph]]; P. Bandyopadhyay, S. Choubey and M. Mitra, JHEP **0910**, 012 (2009) [arXiv:0906.5330 [hep-ph]]; P. Bandyopadhyay, S. Choi, E. J. Chun and K. Min, Phys. Rev. D **85**, 073013 (2012) [arXiv:1112.3080 [hep-ph]].

[5] C. -X. Yue, H. -L. Feng and W. Ma, Chin. Phys. Lett. **27**, 011202 (2010).

[6] A. Abada, C. Biggio, F. Bonnet, M. B. Gavela and T. Hambye, Phys. Rev. D **78**, 033007 (2008) [arXiv:0803.0481 [hep-ph]]; C. Biggio and F. Bonnet, Eur. Phys. J. C **72**, 1899 (2012) [arXiv:1107.3463 [hep-ph]].

[7] S. Chatrchyan *et al.* [CMS Collaboration], Phys. Lett. B **718**, 348 (2012) [arXiv:1210.1797 [hep-ex]]; [ATLAS Collaboration], ATLAS CONF 2013 019; S. Vanini, CERN-THESIS-2012-175. [The CMS Collaboration],[arXiv:1210.1797v2[hep-ex]].

Resolving octant degeneracy at LBL experiment by combining Daya Bay reactor setup

Kalpana Bora[1] **and Debajyoti Dutta**[2]

Physics Department, Gauhati University, Guwahati 781014, Assam, India.

E-mail: [1]kalpana@gauhati.ac.in, [2]debajyotidutta1985@gmail.com

Abstract. Long baseline Experiment(LBL) have promised to be a very powerful experimental setup to study various issues related to Neutrinos. Some ongoing and planned LBL and medium baseline experiments are- T2K, MINOS, NOvA, LBNE, LBNO etc. But, the long baseline experiments are crippled due to presence of some parameter degeneracies, like the Octant - degeneracy. In this work, we first show the presence of Octant degeneracy in LBL experiments and then combine it with Daya Bay Reactor experiment at different values of CP violation phase. We show that the Octant degeneracy in LBNE can be resolved completely with this proposal.

1. Introduction

In the quest to find the unknowns in neutrino sector, long baseline neutrino experiments (LBNEs) have an important role to play. But there is a major drawback of the LBNEs-the presence of parameter degeneracies [1, 2]. The appearance of several disconnected region in multi-dimensional space of oscillation parameters for a given experiment is due to the inherent structure of three flavour neutrino oscillation probabilities. Because of these, it is not exactly possible to pin-point, which one is the exact (true) solution. These degeneracies can be classified in three ways. Appearance of two disconnected regions in the $(\delta_{cp}, \theta_{13})$ plane for $\nu_\mu \to \nu_e$ channel in neutrino and anti neutrino mode leads to intrinsic or $(\delta_{cp}, \theta_{13})$-degeneracy [3, 4]. Appearance of two solutions corresponding to two sign of Δm^2 gives hierarchy or sign (Δm^2)-degeneracy [5] and appearance of solutions corresponding to θ_{23} and $\pi/2-\theta_{23}$ at different values of δ_{cp} and θ_{13} is called octant or (θ_{23}) -degeneracy [6]. In literature, various methods to resolve these degeneracies have been proposed. The use of spectral information [7], combining experiments at various baselines and/or (L/E)-values [2], combining $\nu_\mu \to \nu_e$ and $\nu_e \to \nu_\tau$ oscillation channels [8], combining long baseline(LBL) and reactor experiments [9, 10] and combination of LBNE and atmospheric experiments [11, 12] are a few of them.

Recent global fit of all neutrino oscillation parameters [13], points towards the lower octant $[\theta_{23} < \pi/4]$, but its still not clear which octant is the true octant. Again, bounds on sterile oscillation parameters from Reactor experiments can be found on [14]. In this work, we have considered a LBNE like setup and combined it with Daya Bay [15] reactor setup with the help of simulation using Globes [16] to check what information we can get from it. We call it LBNE-like setup because LBNE collaboration is planning to reconfigure their setup with liquid argon detector. But still our work is relevant in the sense that this result can be implemented to any long baseline experiment having water Cherenkov detector at a baseline $\sim$ 1300 km. LBNE neutrino beamline is driven by the physics of different oscillation channels like -$\nu_\mu \to \nu_e$, $\nu_\mu \to \nu_\mu$, $\nu_\mu \to \nu_\tau$ etc. We have considered the $\nu_\mu \to \nu_e$ appearance channel in this study.

Presence of octant degeneracy, in LBNE like setup is confirmed by exclusion curves in Fig.1(1st column) in the $\theta_{13} - \delta_{CP}$ plane. Next, we combine LBNE like setup with Daya Bay reactor experiment, to resolve the above mentioned octant degeneracy as shown in Fig.1 (2nd column). We find that after combining, the degenerate exclusion curves disappear. With this proposed combination, it is possible to resolve octant degeneracy. In fig.2, we have pinpointed the octant after combining LBNE like setup with Reactor setup at δ_{CP} =0°. We have done this analysis for a running time of 5 years in neutrino and 5 years of anti- neutrino. The results have been presented for different values of δ_{CP} phase $= 0°, 90°, 150°$. We have tested these results for a wide range of δ_{CP}, as we do not know the exact value of this parameter in leptonic sector, from experiments.

2. Experiment details

In our analysis, we have considered the LBNE to DUSEL beamline [17] of 1300 km baseline that is pointing from NuMI (Neutrino Main Injector). The Detector is a 300 kiloton water Cherenkov detector. Running time is 5 years for neutrinos and 5 years for anti-neutrinos. Power considered is 700 KW. Daya Bay [15] has three pairs of reactors at Daya Bay and Ling Ao I, which generate 11.6 GW of power. Daya Bay consists of three underground experimental halls, one far and two near, linked by horizontal tunnels. Eight identical cylindrical detectors are employed to measure the neutrino flux. The mass of each detector is about 20 tons. Equal mass of near and far detector will reduce the systematic errors. Four of these eight detectors are at the far zone while two detectors are kept in each near zone. The distance of the detectors from the reactor cores at the Daya Bay site is 363 m while this distance at the Ling Ao site is 481 m. The far detectors are at 1985 m and 1615 m respectively from the Daya Bay and Ling Ao reactor sites. Energy resolution used is 5%. In this work, we have used only two pairs of reactors.

3. Results and analysis

Using above details of LBNE experiment, we have produced exclusion curves in the $\delta_{CP} - \theta_{13}$ plane. The best fit values of oscillation parameters [13] used are:$\sin^2\theta_{12} = 0.307$, $\sin^2\theta_{23} = 0.427$,$\sin^2\theta_{13} = 0.0241$ and $\Delta m^2_{21} = 7.54 \times 10^{-5} eV^2$ and $\Delta m^2_{31} = 2.43 \times 10^{-3} eV^2$.

Energy resolution used is 10% for the electrons, and 5% for the muons. 1% systematic errors for the signal, and 5% systematic error for the background has been used. While searching for degenerate curves, we give the global best-fit values (true hierarchy, true octant) as true values. Then a chi-square analysis is done, to search for true hierarchy, and wrong octant for the minima of χ^2 function. We have used the $P_{\mu e}$ channel in our calculation and we have marginalised over δ_{CP}, $sin^2 2\theta_{13}$, Δm^2_{31} and θ_{23} in their 3σ ranges. We are only showing plots for normal hierarchy cases. In the plots we abbreviated true hierarchy- true octant as THTO, true hierarchy- wrong octant as THWO.

From these figures, following observations are in order :

(i) Appearance of THTO and THWO (true hierarchy and wrong octant) contours in Fig.1 (1st column, top six panels), indicates the presence of octant degeneracy in the LBNE like experimental setup.

(ii) It may be noted that the regions inside the contours are the allowed regions of the neutrino oscillation parameters θ_{13} and δ_{CP}.

(iii) Occurrence of THTO and THWO curves nearly at same δ_{cp} affects the measurement of θ_{13} but not of δ_{cp}. It means that we can pinpoint the value of δ_{cp} even in the presence of octant degeneracy.

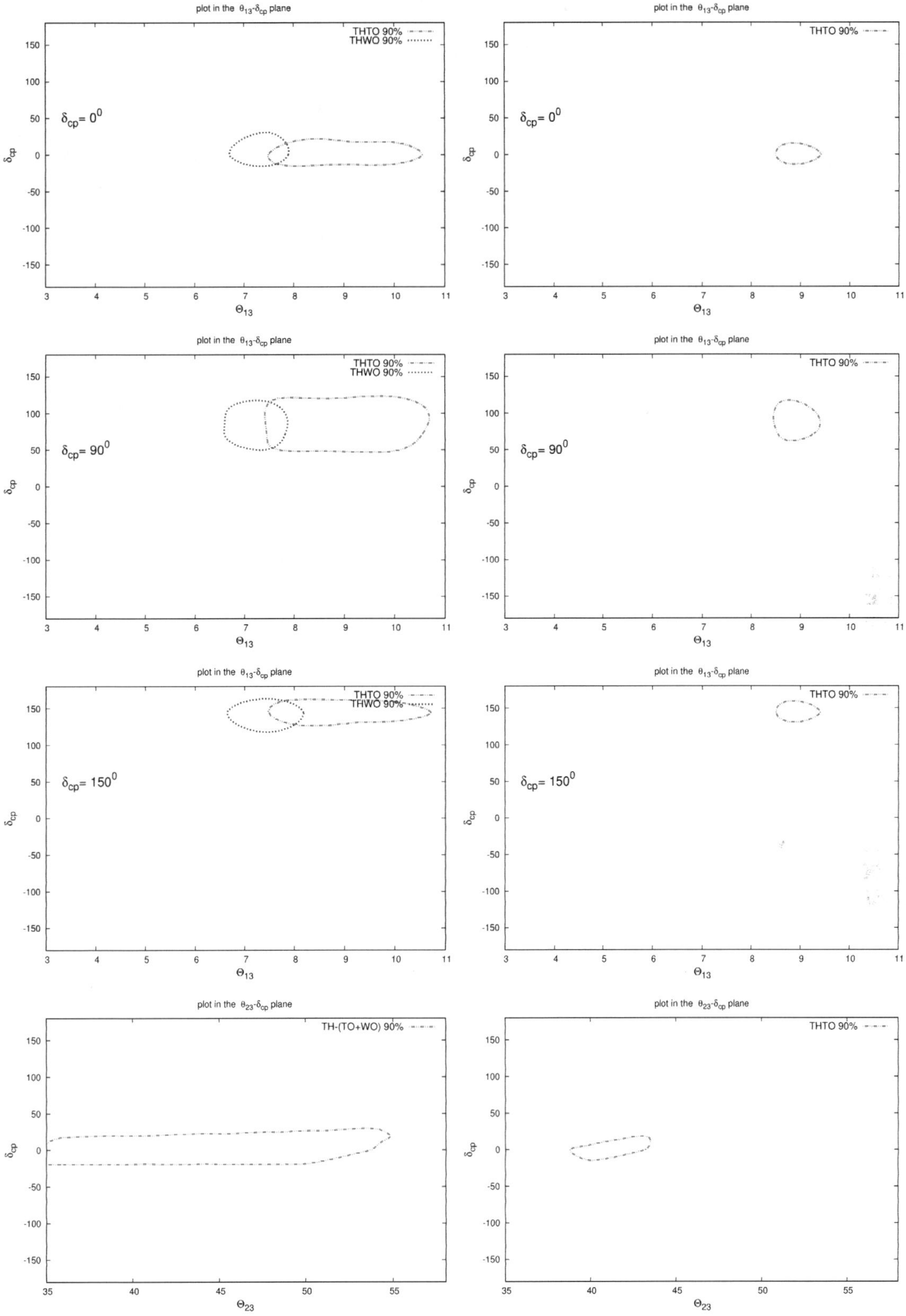

Figure 1. Contours in $\delta_{CP} - \theta_{13}$ plane (top six panels) for LBNE-like setup (1st column) and LBNE-like+Daya Bay setup (2nd column) at different δ_{cp} values and at 90% cl for normal hierarchy. Bottom two panels are for $\delta_{CP} - \theta_{23}$ at $\delta_{cp} = 0$.

(iv) In Fig.1 (2nd column, top six panels), we see that THWO curves disappear, i.e. the Octant degeneracy has been resolved completely, after combining LBNE with Daya Bay experiment. Also we find that the size of the allowed regions decrease, i.e. the measurements of the neutrino oscillation parameters θ_{13} and δ_{CP} become more deterministic. In our opinion, this resolution of Octant degeneracy is because of the fact that, the difference in the values of θ_{13}, for the true and Octant degenerate curves, is more than the sensitivity of the Daya Bay reactor experiment. And therefore, the reactor data can pick up the true value of θ_{23}.

(v) In Fig.1 (bottom two panels, 1st fig), the octant sensitivity is shown in $\delta_{CP} - \theta_{23}$ plane for LBNE like setup at $\delta_{CP} = 0$ degree . It is seen that the contour is spanning over both the octant (true and wrong). Hence from this contour, it is not possible to figure it out which one is the correct octant. But in 2nd fig (bottom panel), which is for the LBNE-like+Daya Bay setup, ambiguity of octant is resolved. It is clear that lower octant ($\theta_{23} < 45^0$) is the true octant as indicted by the disappearance of allowed region in the wrong octant or higher octant of θ_{23}. We have checked the results for different δ_{CP} and the results are consistent. Since, lower octant is also preferred by Global fit data, with the help of this proposal, we have been able to pinpoint the octant in LBNE which is the highlight of this work.

4. Discussion and conclusion

To conclude, in this work, we have presented interesting results on resolution of Octant degeneracy in LBNE-like setup. The Octant degeneracy is resolved after combining LBNE-like with Daya Bay reactor experiment. We can conclude that, in presence Daya Bay reactor setup can pinpoint the true octant of LBNE like setup. Since Octant degeneracy poses a major hurdle to precise measurement of neutrino oscillation parameter θ_{23} at long baseline neutrino experiments, results presented in this work are extremely important and relevant.

5. Acknowledgement

KB would like to thank Thomas Schwetz, Raj Gandhi, and Pomita Ghoshal for fruitful discussions. She also would like to thank UGC-SAP program, and HRI (Harish Chandra Research Institute), Allahabad, for financial support, to visit HRI. DD would like to thank HRI, Allahabad for financial support to visit the institute.

References

[1] K. Bora, JASS, vol. 52,No 2, P 84, 2011 arXiv:hep-ph/1111.7085.
[2] V. Barger, D. Marfatia, K. Whisnant, Phs. Rev D65, 073023 (2002), hep-ph/0112119.
[3] M. Koike, T. Ota, J. Sato, Phys. Rev D65, 053015 (2002), hep-ph/0011387.
[4] J. Burguet-Castell et. al, Nucl. Phys. B608, 301 (2001), hep-ph/0103258.
[5] H. Minakata and H. Nunokawa, JHEP 10, 001 (2001), hep-ph/0108085.
[6] G. L. Fogli, E. Lisi, Phys. Rev D54, 3667 (1996), hep-ph/9604415.
[7] A. Donini, D. Meloni, S. Rigolin, JHEP 06, 011 (2004), hep-ph/0312072.
[8] A. Donini, D. Meloni, P. Migliozzi, Nucl. Phys. B646, 321 (2002), hep-ph/0206034.
[9] H. Minakata and H. Sugiyama, Phys. Lett. B580, 216 (2004), hep-ph/0309323.
[10] P. Huber et. al., Phys. Rev D70, 073014 (2004), hep-ph/0403068.
[11] P. Huber, M. Maltoni, T. Schwetz, hep-ph/0501037.
[12] J.-E. Campagne et. al., hep-ph/0603172.
[13] G. Fogli, E. Lisi, A. Marrone, D. Montanino, A. Palazzo, et al., Phys.Rev.D86 (2012) 013012, [arXiv:1205.5254].
[14] K. Bora, D. Dutta, P. Ghoshal; JHEP 12 (2012) 025; arXiv:1206.2172[hep-ph].
[15] F. P. An et. al. (Daya Bay), Phys. Rev. Lett. 108, 171803 (2012); hep-ex/1203.1669.
[16] P. Huber, J. Kopp, M. Lindner, M. Rolinec, and W. Winter, arXiv:hep-ph/0701187.
[17] T. Akiri, et. al. , arXiv:hep-ph/1110.6249.

Journal of Physics: Conference Series **481** (2014) 012020 doi:10.1088/1742-6596/481/1/012020

An analysis of Isgur-Wise function of heavy-light mesons within a higher dimensional potential model approach

Sabyasachi Roy[1] and D K Choudhury[1,2,3]
[1]Department of Physics, Gauhati University, Guwahati 781014, India.
[2]Centre for Theoretical Studies, Pandu College, Guwahati 781012, India
[3]Physics Academy of the North East, Department of Physics, Gauhati University, Guwahati 781014, India.

Abstract. Nambu-Goto action for bosonic string predicts the quark-antiquark potential to be $V(r) = -\gamma/r + \sigma r + \mu_0$. The coefficient $\gamma = \pi(d-2)/24$ is the Lüscher coefficient of the Lüscher term γ/r, which depends upon the space-time dimension 'd'. Very recently, we have developed meson wave functions in higher dimension with this potential from higher dimensional Schrödinger equation by applying quantum mechanical perturbation technique with both Lüscher term as parent and as perturbation. In this letter, we analyze Isgur-Wise function for heavy-light mesons using these wave functions in higher dimension and make a comparative study on the status of the perturbation technique in both the cases.
Key words : Nambu-Goto potential, Lüscher Term, Airy's function, space-time dimension.
PACS Nos. : 12.39.-x , 12.39.Jh , 12.39.Pn, 03.65.Ge.

1. Introduction

In the non-perturbative low energy regime of QCD theory, potential model approach has been successful, even in the non-perturbative approximations, for the study of quark-antiquark bound states. There are several potentials in literature for modeling mesons like Cornell potential [1], Martin potential [2], Logarithmic potential [3], Richardson potential [4] etc. Out of these, the *linear plus Coulombic* type Cornell potential is believed to be the more realistic one to account for quark-antiquark interaction in a meson, as it includes both the QCD concepts of asymptotic freedom (Coulombic term) and confinement (linear term).

$$V(r) = -\frac{4\alpha_s}{3r} + br + c \tag{1}$$

In the non-relativistic approach, two-body Schrödinger equation is employed with such a potential to extract meson wave function and subsequently analysis on static and dynamic properties of mesons are made. The main problem lying within such approach is that Schrödinger equation is not exactly solvable with such a potential. In recent past, we have successfully applied quantum mechanical perturbation technique in solving Schrödinger equation with such a

[1] Corresponding author. On leave from Karimganj College, Assam, India. e-mail : *sroy.phys@gmail.com*

Table 1. ρ^2 and C (with Lüscher parent).

	D meson		B meson	
D	ρ^2	C	ρ^2	C
3	186.1	11200.7	186.4	11208.8
4	76.4	2032.1	76.9	2051.6
5	50.51	854.9	50.82	563.7
9	29.39	204.2	29.53	208.8
15	26.93	165.4	27.21	169.2
20	22.41	96.6	22.68	99.2
25	18.83	64.23	19.02	64.6
∞	15.32	36.64	15.55	36.98

Table 2. Standard ρ^2 and C values.

Model / collaboration	ρ^2	C
Ref [16]	1.8	0.95
Skryme Model [17]	1.3	0.85
Neubert [18]	0.82	–
UK QCD Collab. [19]	0.83	–
CLEO [20]	1.67	–
BELLE [21]	1.35	–
HFAG [22]	1.17	–
Huang [23]	1.35	–

potential. Very recently, following Nambu-Goto action for bosonic strings [7], we have introduced [5,6] higher dimensional potential model in our studies on heavy-light mesons which is

$$V(r) = -\frac{\gamma}{r} + \sigma r + \mu_0 \tag{2}$$

Here, $\gamma = -\pi(d-2)/24$ is the Lüscher coefficient [8,9] of the Lüscher term γ/r, which depends upon the space-time dimension 'd' and σ is the string tension whose generally accepted value is $0.89\,\mathrm{GeV/fm}$; μ_0 is a regularisation dependent mass term. We have employed Schrödinger equation in higher dimension and develop meson wave function in both the cases of perturbation - Lüscher term as parent [5] and linear term as parent [6]. Then, we have made subsequent studies on Isgur-Wise function (IWF) of heavy-light mesons in both the cases.

In this work, we make some refinement of our analysis of IWF in higher dimension and put forward our comment regarding the status of perturbation for both the choices of 'parent and child', within such higher dimensional potential model.

2. Theory and Calculation

2.1. With Lüscher term as parent

With Lüscher term as parent, the wave function is obtained as [5],

$$\Psi^{\text{total}}(r) = N_1\left[1 - \frac{\sigma(2D-3)}{6\gamma}r^2\right] r^{\frac{D-3}{2}} e^{-\mu\gamma r} \tag{3}$$

where D is the spatial dimension such that $d = D+1$. N_1 is the normalisation constant which is obtained from

$$\int_0^\infty DC_D r^{D-1}\left|\Psi^{\text{total}}(r)\right|^2 dr = 1 \tag{4}$$

This gives

$$N_1 = \frac{1}{(DC_D)^{1/2}}\left[\frac{\Gamma(2D-3)}{(2\mu\gamma)^{2D-3}} - 2k\frac{\Gamma(2D-1)}{(2\mu\gamma)^{2D-1}} + k^2\frac{\Gamma(2D+1)}{(2\mu\gamma)^{2D+1}}\right]^{-1/2}. \tag{5}$$

Here, we consider $k = \sigma(2D-3)/(6\gamma)$ and $C_D = \pi^{D/2}/\Gamma(D/2+1)$ such that $d\tau = DC_D r^{D-1}dr$ represents volume element in D dimensional spherical coordinates[10]. It is to be mentioned here that at $D = 3$, this higher dimensional wave function gives back our three-dimensional result

Table 3. ρ^2 and C for different D values (with linear parent).

	D meson		B meson	
D	ρ^2	C	ρ^2	C
3	0.2158	0.0174	0.2608	0.0254
4	0.3874	0.0457	0.4039	0.0483
5	0.4366	0.0613	0.4813	0.0679
9	0.8355	0.1746	0.9497	0.1916
15	0.9622	0.2710	1.0281	0.3274
20	1.2178	0.3997	1.3352	0.4887
25	1.3732	0.5041	1.4637	0.6623

[11], when γ is replaced by $4\alpha_s/3$ and σ by b. Considering volume element in D dimension, the derivatives of IWF are obtained from

$$\rho^2 = DC_D\mu^2 \int_0^\infty r^{D+1} \left|\Psi^{\rm total}(r)\right|^2 dr, \tag{6}$$

$$C = \frac{1}{6}DC_D\mu^4 \int_0^\infty r^{D+3} \left|\Psi^{\rm total}(r)\right|^2 dr, \tag{7}$$

With wave function given in (3), we obtain explicit expressions of ρ^2 and C,

$$\rho^2 = \mu^2 \frac{\left[\frac{\Gamma(2D-1)}{(2\mu\gamma)^{2D-1}} - 2k\frac{\Gamma(2D+1)}{(2\mu\gamma)^{2D+1}} + k^2\frac{\Gamma(2D+3)}{(2\mu\gamma)^{2D+3}}\right]}{\left[\frac{\Gamma(2D-3)}{(2\mu\gamma)^{2D-3}} - 2k\frac{\Gamma(2D-1)}{(2\mu\gamma)^{2D-1}} + k^2\frac{\Gamma(2D+1)}{(2\mu\gamma)^{2D+1}}\right]}, \tag{8}$$

$$C = \frac{\mu^4}{6} \frac{\left[\frac{\Gamma(2D+1)}{(2\mu\gamma)^{2D+1}} - 2k\frac{\Gamma(2D+3)}{(2\mu\gamma)^{2D+3}} + k^2\frac{\Gamma(2D+5)}{(2\mu\gamma)^{2D+5}}\right]}{\left[\frac{\Gamma(2D-3)}{(2\mu\gamma)^{2D-3}} - 2k\frac{\Gamma(2D-1)}{(2\mu\gamma)^{2D-1}} + k^2\frac{\Gamma(2D+1)}{(2\mu\gamma)^{2D+1}}\right]}. \tag{9}$$

The values of ρ^2 and C for different D values are reported in Table-1. For comparison, some recent standard theoretical and experimental results are shown in Table-2.

2.2. With linear term as parent

Considering linear term in potential as parent in perturbation method, the wave function comes out as

$$\Psi^{\rm total}(r) = N_1 r^{\frac{(1-D)}{2}} \left[1 + A_1(r,D)r + A_2(r,D)r^2 + A_3(r,D)r^3 + \cdots\right] (\varrho_1 r)^m Ai[\varrho_1 r - \varrho_0]. \tag{10}$$

Terms A_1, A_2 etc, Airy function $Ai[\varrho]$ and other terms involved in wave function are described in ref [6]. Normalisation constant N_1 and slope (ρ^2) and curvature (C) can be obtained using equations (4), (6) and (7) respectively. To avoid divergences due to infinite upper limit of integration in these expressions, from convergence condition of perturbation series, we have earlier introduced reasonable cut-off value r_0, which is obviously dimension dependent (Table-1 of ref[6]). The variation of ρ^2 and C with dimension D is reported in Table-3 of ref [6], which we also report here as Table-3, for convenience of our analysis.

3. Results and discussion

In the case of Lüscher term as parent, although our results [Eqs.(3,8,9) for $D = 3$ give back the corresponding expressions obtained with Cornell potential [11]) when γ is replaced by $4\alpha_s/3$ and σ by b, still our result for ρ^2 and C (Table-1) are higher than the standard values.

However, at higher D, when Lüscher term becomes more and more dominant, the values of ρ^2 and C go on decreasing, the asymptotic limits being $\rho^2_{\text{asym}} = 15.32$ and $C = 36.64$ for D meson. Here, we mention that, we have worked with fixed value of confinement parameter($\sigma = 0.89\,\text{GeV/fm}$) , unlike earlier analysis with Cornell potential [11], where confinement parameter b is varied suitably to meet the expectations. Also, the Coulomb-type Lüscher term depends only upon dimensional parameter D, where as in Cornell potential the coefficient of Coulombic term contains strong coupling constant α_s, bringing in flexibility in the calculations.

On the other hand, with linear term as parent in perturbation technique, our results (Table-3) are reasonably close to the standard values (Table-2). More importantly, in this case, ρ^2 and C values increase with increase in D, although remaining within the range of our expectations. We have restricted our calculation up to the term $A_5(r, D)$ in the first infinite series of the wave function; there remains scope for further refinement of results.

From our analysis of derivatives of IWF of heavy-light mesons in both the choices of 'parent-child', we can infer that the case of linear confinement term as parent is giving more reliable results compared to the case of Lüscher term as parent. This is also supported by recent lattice calculation [14]. Also, theoretically it is reasonable that the higher dimensional potential originates from a more general string inspired exact potential enunciated by J. F. Arvis [15].

$$V(r) = \sigma r\sqrt{1 - \frac{\pi(d-2)}{12\sigma r^2}} \tag{11}$$

On expansion, it gives linear term as leading one with Lüscher term as first order correction to it. This supports the case of treating linear term as parent with Lüscher term as first order perturbation to it.

Acknowledgement
SR acknowledges the support of University Grants Commission, Govt. of India in terms of fellowship under FDP scheme to pursue his research work at Gauhati University.

References

[1] Eicheten E *et al.* 1978, *Phys. Rev.* D **17** 3090; Eicheten E *et al.* 1980 *Phys. Rev.* D **21** 203.
[2] Martin A 1980 *Phys. Lett.* B **93**338.
[3] Quigg C and Rosuer J L 1977 *Phys. Lett.* B **71** 153.
[4] Richardson H D 1979 *Phys. Lett.* B **82** 272.
[5] Roy S , Hazarika B J and Choudhury D K 2012 *Phys. Scr.* **86** 045101.
[6] Roy S and Choudhury D K 2013 *Phys. Scr.* **87** 065101.
[7] Nambu Y 1979 *Phys. Lett.* B **80** 372.
[8] Lüscher M, Symanzik K and Weisz P 1980 *Nucl.Phys.* B **173** 365.
[9] Lüscher M 1981 *Nucl. Phys.* B **180** 317.
[10] Eduarda M, Henderson David G 1996 in "Experiencing Geometry: On plane and sphere", Prentice Hall.
[11] Bordoloi N S and Choudhury D K 2000 *IJMP* A **23** 3667.
[12] Isgur N and Wise M B 1991 *Phys. Rev. Lett.* **66** 1130.
[13] Roy S, Bordoloi N S and Choudhury D K 2013 *Can.J.Phys* **91** 34.
[14] Bali G S 1999 *Phys. Lett.* B **460** 170 [hep-ph/9905387].
[15] Arvis J F 1983 *Phys. Lett.* B **127** 106.
[16] Roy S and Choudhury D K 2012 *Mod. Phys. Lett.* A **27** 1250110.
[17] Jenkins E, Manahar A, Wise M B 1996 *Nucl. Phys* B **38** 396.
[18] Neubert M 1991 *Phys Lett* B **204** 455.
[19] UKQCD Collab. : Bowler K C *et al.* 2002 *Nucl Phys* B **637** 293.
[20] CLEO Collab. : Bartel J *et al.* 1999 Phys Rev Lett **82** 3746.
[21] BELLE collab. : Abe K *et al.* 2002 *Phys Lett* B **526** 258.
[22] Heavy Flavor Averaging Group (HFAG) 2009 hep-ex/08081297.
[23] Huang Ming-Qiu *et al.* 2005 *Phys. Letts.* B **629** 27.

An improved analysis of neutrino-nucleon interaction cross-section at ultra high energy

Pijush Kanti Dhar [1] **and D K Choudhury** [2]

[1] Department of Physics, Arya Vidyapeeth College, Guwahati 781016, India.
[2] Department of Physics, Gauhati University, Guwahati 781014, India.

E-mail: dr_pijush@yahoo.co.in

Abstract. We determine the Ultra High Energy (UHE) neutrino-nucleon interaction cross-sections numerically both for the charged and the neutral current in the leading order (LO) and find our numerically determined results to be in sufficient agreement with the results obtained by other authors. For this, we exclusively carry out extrapolation at ultra low x and find our method to be satisfactory.

1. Introduction

In this paper, cross-sections of UHE neutrino-nucleon interaction are determined by carrying out numerical integration of the double differential cross-section at one-loop level. For this purpose, we take the help of GRV94 parton distributions [1]. We then compare our predictions with the results of several other authors [2–4] and our previously done analytical results [5] obtained by using the solutions of the non-singlet and singlet Dokshitzer-Gribov-Lipatov-Altarelli-Parisi (DGLAP) equations respectively [6].

2. Formalism

2.1. Charged current-numerical solution

The differential cross-section for the charged current (CC) UHE interaction is given by [7],

$$\frac{d^2\sigma_{CC}^{\nu(\overline{\nu})N}}{dxdQ^2} = \frac{G_F^2}{2\pi}\left(1+\frac{Q^2}{M_W^2}\right)^{-2}\frac{1}{x}\left[\left(1-y+\frac{y^2}{2}\right)F_2^{\nu(\overline{\nu})^{CC}}(x,t) \right.$$
$$\left. \pm y(1-\frac{y}{2})xF_3^{\nu(\overline{\nu})^{CC}}(x,t)\right]. \quad (1)$$

where Fermi coupling constant $G_F = 1.1663 \times 10^{-5}$ GeV^{-2}, $Q^2 = xys$, $s = 2ME_{\nu(\overline{\nu})}$ is the centre-of-mass energy squared, x is the usual Bjorken variable, y is the inelasticity parameter. The total cross-section for UHE CC interaction in an isoscaler target $N[\equiv \frac{(p+n)}{2}]$ is given by,

$$\sigma_{CC}^{\nu(\overline{\nu})N} = \int_{\frac{Q^2}{s}}^{1} dx \int_{Q_0^2}^{s} dQ^2 \frac{d^2\sigma_{CC}^{\nu(\overline{\nu})N}}{dxdQ^2}. \quad (2)$$

We only consider all light (u, d, s) quarks. We have the following LO expressions in QCD for charged current (CC) interactions [7],

$$F_1^{\nu,light} = \frac{1}{2}\left(\overline{u}+\overline{d}\right) + \frac{1}{2}(d+u)|V_{ud}|^2 + s|V_{us}|^2. \tag{3}$$

$$F_2^{\nu,light} = 2xF_1^{\nu,light} \tag{4}$$

$$F_3^{\nu,light} = -\left(\overline{u}+\overline{d}\right) + (d+u)|V_{ud}|^2 + 2s|V_{us}|^2. \tag{5}$$

where $u(x,t)$, $d(x,t)$ etc. represent parton distributions at various values of x and t and V_{ud}, V_{us} etc. are the relevant CKM matrix elements [8]. For obtaining the expression of total cross-section, we first put the explicit parton distributions equations (3-5) in equation (1) and then with the help of equation (2), we finally get [9]

$$\begin{aligned}\sigma_{CC}^{\nu(\overline{\nu})N} &= \frac{G_F^2}{2\pi}\int_{Q_0^2}^{s} dQ^2\left(1+\frac{Q^2}{M_W^2}\right)^{-2}\int_{\frac{Q^2}{s}}^{1}\frac{dx}{x}\left[\left(1-\frac{Q^2}{xs}+\frac{Q^4}{2x^2s^2}\right)2x\left\{\frac{1}{2}\left[\overline{u}(x,t)\right.\right.\right.\\ &\left.+\overline{d}(x,t)\right]+\frac{1}{2}[d(x,t)+u(x,t)]|V_{ud}|^2+s|V_{us}|^2\Big\}\pm\frac{Q^2}{xs}\left(1-\frac{Q^2}{2xs}\right)x\\ &\left.\left\{-\left[\overline{u}(x,t)+\overline{d}(x,t)\right]+[d(x,t)+u(x,t)]|V_{ud}|^2+2s|V_{us}|^2\right\}\right].\end{aligned} \tag{6}$$

2.2. Neutral current-numerical solution

The total cross-section for UHE NC interaction (σ_{NC}) in an isoscalar target $N[\equiv \frac{(p+n)}{2}]$ is similar to equation (2), viz.

$$\sigma_{NC}^{\nu(\overline{\nu})N} = \int_{\frac{Q^2}{s}}^{1} dx \int_{Q_0^2}^{s} dQ^2 \frac{d^2\sigma_{NC}^{\nu(\overline{\nu})N}}{dxdQ^2}. \tag{7}$$

where

$$\begin{aligned}\frac{d^2\sigma_{NC}^{\nu(\overline{\nu})N}}{dxdQ^2} &= \frac{G_F^2}{2\pi}\left(1+\frac{Q^2}{M_Z^2}\right)^{-2}\frac{1}{x}\left[\left(1-y+\frac{y^2}{2}\right)F_2^{\nu(\overline{\nu})^{NC}}(x,t)\right.\\ &\left.\pm y\left(1-\frac{y}{2}\right)xF_3^{\nu(\overline{\nu})^{NC}}(x,t)\right].\end{aligned} \tag{8}$$

Putting the LO expressions for structure functions involving Neutral currents (NC) [7] for light quarks in equation (8) and then with the help of equation (7), we get the expression of total cross-section for neutral current as [9]

$$\begin{aligned}\sigma_{\mathrm{NC}}^{\nu(\overline{\nu})N} &= \frac{G_F^2}{2\pi}\int_{Q_0^2}^{s} dQ^2\left(1+\frac{Q^2}{M_Z^2}\right)^{-2}\int_{\frac{Q^2}{s}}^{1}\frac{dx}{x}\left[\left(1-\frac{Q^2}{xs}+\frac{Q^4}{2x^2s^2}\right)\{0.1432x\right.\\ &\times\left[u(x,t)+\overline{u}(x,t)+d(x,t)+\overline{d}(x,t)\right]+0.1849x\\ &\times\left[u(x,t)+\overline{u}(x,t)+d(x,t)+\overline{d}(x,t)+4s\right]\}\\ &\left.\pm\frac{Q^2}{xs}\left(1-\frac{Q^2}{2xs}\right)\left\{0.268x\left[u(x,t)-\overline{u}(x,t)+d(x,t)-\overline{d}(x,t)\right]\right\}\right]\end{aligned} \tag{9}$$

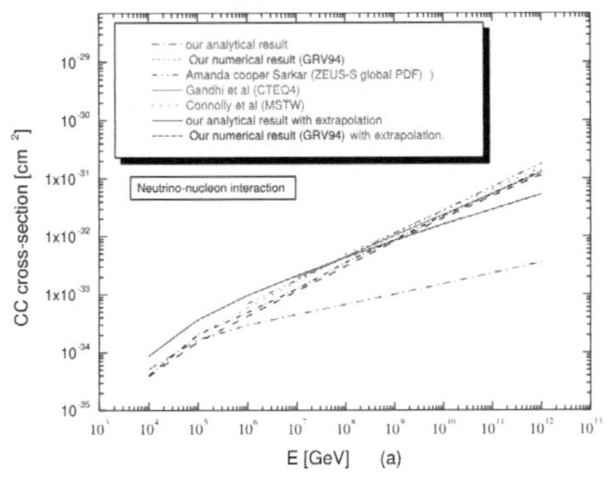

Figure 1. Numerically obtained cross-sections $\sigma_{CC}^{\nu N}\,(cm^2)$ are compared with other results [2–5].

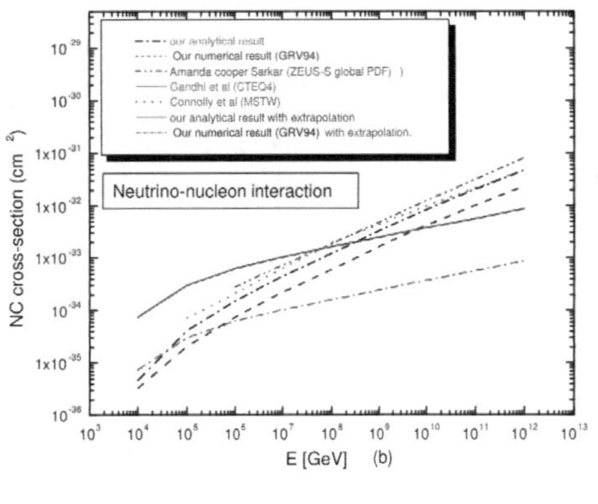

Figure 2. Numerically obtained cross-sections $\sigma_{NC}^{\nu N}\,(cm^2)$ are compared with other results [2–5].

2.3. Extrapolation at ultra low x

Since the UHE neutrinos coming from various sources have energy more than 10^5 GeV, so perturbative calculations should be carried out using parton distribution functions at Ultra-low Bjorken x ($x < 10^{-5}$). But high energy measurements of deep inelastic scattering (DIS) of lepton-nucleon at the collider HERA [10] at DESY provide data for ($x \geq 10^{-5}$). So we carry out extensive extrapolation in the region of x still unexplored by terrestrial experiments. This is done by multiplying $F_2(x,t)$ by $\log(\frac{1}{x})$ in equation (1) and equation (8), consequently modifying equation (6) and equation (9) for charged and neutral currents respectively. It is worthwhile to mention that the contribution from xF_3 to cross-section is negligible at Ultra High Energy of neutrino.

3. Results and discussions

We use GRV94 generalised parton distributions [1] for determining the cross-section numerically. Putting these distributions in equation (6) and equation (9), we obtain the values of non-extrapolated cross-sections for νN interactions both for the charged current and the neutral current respectively. For getting extrapolated numerical results, we follow the techniques as mentioned in Subsection 2.3.

In Figure 1, we plot our numerical LO results (with and without extrapolation) for $\sigma_{CC}^{\nu N}\,(cm^2)$ versus neutrino energy. In the same Figure, we also plot next-to-leading order (NLO) results of various other authors [2–4]. Similarly in Figure 2, we compare our LO predictions for $\sigma_{NC}^{\nu N}\,(cm^2)$ (with and without extrapolation) with the NLO results of various other authors [2–4]. In both the cases, we find that our numerically-extrapolated results match better with the result of other authors than our numerically non-extrapolated results. Due to the absence of propagator effect in low energy regime, our numerical result indicates a sharp rise for $E_\nu \leq 10^3$ GeV. For $E_\nu \geq 10^3$ GeV, propagator effect creates the dampening. But there is a severe dampening in our previously calculated non-extrapolated analytical results [5] for $E_\nu \geq 10^3$ GeV, much more than our non-extrapolated numerical results. Of course, our extrapolated analytical results [5] agree more or less well with our extrapolated numerical results and the results of various other authors [2–4] as clearly seen from Figure 1 and Figure 2.

4. Comments and conclusions

To conclude, we have shown that our numerically-extrapolated result agree quite well with the results of other authors [2–4] both at the lower as well as higher ends of the neutrino energy spectrum. This reflects that the exclusive technique adopted by us in this paper for carrying out extrapolation at ultra low x is quite acceptable and satisfactory. Taking into consideration the contribution due to heavy quarks might improve this result marginally.

References

[1] Glück M, Reya E, and Vögt A, 1995 *Z. Phys.* C **67** 433
[2] Gandhi R, Quigg C, Reno M and Sarcevic I 1998 *Phys. Rev.* D **58** 093009, arXiv: hep-ph/9807264
[3] Connolly A, Thorne Robert S and Waters D, arXiv: 1102.0691
[4] Sarkar A C and Sarkar S, arXiv: 0710.5303
[5] Choudhury D K and Dhar P K, *'Analytical evolution of UHE neutrino-nucleon interaction cross-section'*, sent to *Physics Academy of North-East (PANE) Proceedings Journal*
[6] Altarelli G and Parisi G, 1977 *Nucl. Phys.* B **126** 298
[7] Glück M, Kretzer S and Vögt A, 1999 *Astropart. Phys.* **11** 327, arXiv:astro-ph/9809273
[8] Amsler C *et al*, 2008 *'Review of Paticle Physics: The CKM Quark-mixing angles' Phys. Lett.* B **667** 1
[9] Choudhury D K and Dhar P K, *'Cross-sections of the Neutrino-nucleon Interaction at Ultra-high Energy'*, accepted in *Journal of Korean Physical Society (JKPS)*
[10] H1 Collab: Adloff C *et al.* 2001 *Phys. Lett.* B **520** 183

Comments on the perturbation of Cornell potential in a QCD potential model

D K Choudhury[1] and Krishna Kingkar Pathak[2]

[1]Physics Department, Gauhati University, Guwahati 781014, India.
[2]Physics Department, Arya Vidyapeeth College, Guwahati 781016, India.

E-mail: kkingkar@gmail.com

Abstract. We find in the analysis that the linear part of the Cornell potential can be treated as perturbation for a set of larger values of α_s in the range $0.4 \leq \alpha_s \leq 0.75$ with a constant shift within the range of $-0.4\,\text{GeV} \leq c \leq -1\,\text{GeV}$. Moreover with the same range of constant shift in the Potential, we expect better results with Coulombic part as perturbation for $\alpha_s \leq 0.4$.

1. Introduction

In the potential models, the effective potential between a quark and antiquark can be taken as the Coulomb-plus-linear potential,

$$V(r) = -\frac{4\alpha_s}{r} + br + c. \tag{1}$$

This potential has received a great deal of attention in particle physics, more precisely in the context of meson spectroscopy where it is used to describe systems of quark and antiquark bound states.However, it has been found to be questionable about the numbers of free parameters(α_s, b, c) and numbers of findings in any potential model. The success of a phenomenological model depends on reducing the free model parameters to obtain more precise values with proper arguments and analysis.

In this letter, we put forward the comments on linear part of the Potential as perturbation with Coulombic part as Parent [1, 2] as well as Coulombic part as perturbation with linear as parent [3] in a potential model and attempt to put some constraints on the model parameters.

2. The method of perturbation

It is well known that one cannot solve the Schrdinger equation in quantum mechanics with the QCD potential(equation (1)) except for some simple models. Perturbation theory has been helpful since the earliest applications of quantum mechanics in this regard. In fact, perturbation theory is probably one of the approximate methods that most appeals to intuition [4].
The advantage of taking Cornell Potential for study is that it leads naturally to two choices of "parent" Hamiltonian, one based on the Coulomb part and the other on the linear term, which can be usefully compared. It is expected that a critical role is played by r_0 where the Potential $V(r) = 0$. Aitchison and Dudek in Reference [5] put an argument that if the size of a state measured by $\langle r \rangle < r_0$, then the Coulomb part as the "Parent" will perform better

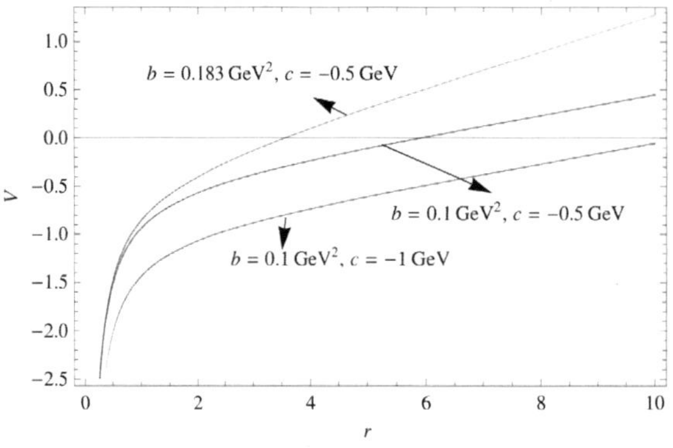

Figure 1: Variation of $V(r)$ with the variation of b, c and α_s.

and if not so the linear part as "parent" will perform better. The Aitchison's work also showed that with Coulombic part as perturbation(VIPT), bottomonium spectra are well explained than charmonium where as charmonium states are well explained with linear part as parent. It becomes noteworthy in this context that the critical distance r_0 is not a constant and can be enhanced by reducing b and c or by increasing α_s. In Figure 1, we show the variation of $V(r)$ with the variation of model parameters.

3. The QCD potential model

For completeness and proper reference we put the last modified version of our model wave function with Coulombic part as parent as [6, 7]

$$\psi_{\text{rel+conf}}(r) = \frac{N'}{\sqrt{\pi a_0^3}} e^{\frac{-r}{a_0}} \left(C' - \frac{\mu b a_0 r^2}{2}\right) \left(\frac{r}{a_0}\right)^{-\epsilon} \tag{2}$$

where N' is the normalisation constant. All other terms involved in equation 2 are explained in reference [1, 2, 6] with a correction for ϵ [7]

$$\epsilon = 1 - \sqrt{1 - \left(\frac{4}{3}\alpha_s\right)^2}. \tag{3}$$

4. Constraints from two points of view

In this section, we discuss the constrains of the model parameters. The values of α_s and the constant shift$c$ of the potential $V(r)$ are expected to fit from mass spectroscopy of hadrons to study its other properties. A narrow range of the free parameters in a potential model measures its success and applicability as well. Here we have tried to show some constrains on the free parameters α_s and c from two points of view.

4.1. From the convergence point of view

In Reference [1, 2], it is shown from the momentum transform of equation(2) (with $C' = 1$) that confinement can be treated as perturbation provided

$$\frac{(4-\epsilon)(3-\epsilon)\mu b a_0^3}{2(1 + a_0^2 Q^2)} \ll 1. \tag{4}$$

Table 1: Q_0^2 (in GeV^2) from Ref. [1]

Mesons	$\alpha_s = 0.65$	$\alpha_s = 0.6$	$\alpha_s = 0.5$
D	3.5	4.1	5.5
D_s	3.5	4.2	5.8
B	1.4	1.7	2.3
B_s	3.9	4.4	5.5
B_c	0	0.4	2.0

Table 2: Q_0^2 (in GeV^2) from equation(5)

Mesons	$\alpha_s = 0.5$	$\alpha_s = 0.4$
D	0.0264	0.0362
D_s	0.0304	0.0490
B	0.0286	0.0412
B_s	0.0300	0.0530
B_c	0	0.0162

Table 3: Values of r_1 and r_0 for $c = -0.5\,\mathrm{GeV}$

mesons	$\alpha_s = 0.65$		$\alpha_s = 0.5$	
	r_1	r_0	r_1	r_0
$D(c\bar{u}/cd)$	6.26	3.94	8.14	3.71
$D(c\bar{s})$	4.70	3.94	6.11	3.71
$B(\bar{b}u/\bar{b}d)$	5.50	3.94	7.15	3.71
$B_s(\bar{b}s)$	3.93	3.94	5.1	3.71
$B_c(\bar{b}c)$	1.46	3.94	1.90	3.71

Table 4: Values of r_1 and r_0 for $c = -1\,\mathrm{GeV}$

mesons	$\alpha_s = 0.65$		$\alpha_s = 0.5$		$\alpha_s = 0.4$	
	r_1	r_0	r_1	r_0	r_1	r_0
$D(c\bar{u}/cd)$	6.26	6.22	8.14	6.06	10.18	5.95
$D(c\bar{s})$	4.70	6.22	6.11	6.06	7.63	5.95
$B(\bar{b}u/\bar{b}d)$	5.50	6.22	7.15	6.06	8.93	5.95
$B_s(\bar{b}s)$	3.93	6.22	5.1	6.06	6.39	5.95
$B_c(\bar{b}c)$	1.46	6.22	1.90	6.06	2.38	5.95

The values of Q_0^2 from equation (6) with $b = 0.183\,\mathrm{GeV}^{-2}$ for B and D mesons are shown in Table 1. The standard spectroscopic result $b = 0.183\,\mathrm{GeV}^2$ [8] can be accommodated by a proper choice of c [2, 6], so that the perturbative condition of equation (4) becomes

$$\frac{(4-\epsilon)(3-\epsilon)\mu b a_0^3}{2(1+a_0^2 Q^2)} \ll C'. \tag{5}$$

With this condition, one can impose $b = 0.183\,\mathrm{GeV}^2$ for low Q^2 value. The improved values of Q_0^2 from equation(5) with $b = 0.183\,\mathrm{GeV}^2$, $m_{u/d} = 0.33\,\mathrm{GeV}$,$m_s = 0.483\,\mathrm{GeV}$,$m_c = 1.55\mathrm{GeV}$,$m_b = 4.93\mathrm{GeV}$ for B and D mesons are presented in Table 2.

Thus to incorporate lower value of Q^2 ($Q^2 \leq \Lambda_{QCD}^2$),with linear part as perturbation in the improved version one expects a bound of α_s as $0.4 \leq \alpha_s \leq 0.75$. The results of the References [6, 9] also indicates that with $\alpha_s = 0.6$ one can expect high accuracy towards experimental and other theoretical values.

4.2. From the condition of Aitchison and Dudek

Considering the argument of Aitchison and Dudek [5] $\langle r \rangle < r_0$ to treat the linear part as perturbation, we get

$$\langle r \rangle_{\mathrm{coul}} = \int \psi^* r \psi dr = \frac{3a_0}{2} = r_1\,(\mathrm{say}) \tag{6}$$

and the critical distance r_0 at which $V(r_0) = 0$ can be obtained by the relation

$$br_0^2 + cr_0 - \frac{4\alpha_s}{3} = 0. \tag{7}$$

The variation of r_1 and r_0 with the model parameters can be easily studied from the above relations. From the calculation it seems to be clear that to treat linear part as perturbation with the valid condition $\langle r\rangle < r_0$ one has to choose the value of $c \ll -0.5$ GeV. With $c = -1$GeV the condition is found to be valid for certain value of α_s. In Table 3 and Table 4, we present our result with $c = -0.5$ GeV and $c = -1$ GeV.

4.3. Constrains on α_s

From the above analysis we see that in the perturbation procedure the value of α_s and the model parameter c plays a crucial role in choosing the parent and perturbative terms. From the reality condition and convergence of series demands the value of α_s within the range of $0.4 \leq \alpha_s \leq 0.75$ in the model without putting any further restriction or constraints. However the logarithmic decrease of α_s depends on the QCD energy scale parameter Λ_{QCD} which is a free parameter and has to be measured in the experiments. One well known formula to fix the value of α_s in Quark models is taken as [7, 10]

$$\alpha_s\left(\mu^2\right) = \frac{4\pi}{\left(11 - \frac{2n_f}{3}\right)\ln\left(\frac{\mu^2+M_B^2}{\Lambda_{QCD}^2}\right)} \tag{8}$$

where, n_f is the number of light flavours, μ is renormalistion scale related to the constituent quark masses as $\mu = 2\frac{m_i m_j}{m_i+m_j}$. M_B is the background mass related to the confinement term of the potential as $M_B = 2.24 \times b^{1/2} = 0.95$ GeV. The reality condition of α_s in equation(8) requires that $\Lambda_{QCD} \leq 460\,\text{MeV}$. By fitting the ρ meson mass in equation(8) one easily obtains QCD scale parameter $\Lambda_{QCD} = 413\,\text{MeV}$ [10].

5. Conclusion and comments

In this letter we mainly devote in finding the analytical conditions to treat the linear part of the Cornell potential as perturbation. We find from the convergence point of view that one can consider the confining part of the potential as perturbation with $0.4 \leq \alpha_s \leq 0.75$ and $c = -0.5$GeV. From table.3 and table.4 it seems to be clear that the validity of the condition $\langle r\rangle < r_0$ demands parametrisation of $\langle c\rangle < -0.5\,\text{GeV}$ and $\alpha_s > 0.6$. However with linear part as perturbation, if the value of α_s in the above range is taken to be granted, then with the same potential another possibility of considering the coulombic part as perturbation also arises for a value of $\alpha_s \leq 0.4$. Interestingly, in Reference. [3], it is shown that with $\alpha_s = 0.39$ and $\alpha_s = 0.22$ one can obtain the required values of slope and curvature in the model with coulombic part as perturbation. The results in Reference [3], clearly indicates that with coulombic part as perturbation, one can get improved results with $\alpha_s \leq 0.4$ than $\alpha_s \geq 0.4$.

References

[1] Choudhury D K *et al*; 1995 *PRAMANA Journal of Physics* **44** 6
[2] Choudhury D K and Bordoloi N S 2002 *Mod. Phys.Lett.* A **17** 1909
[3] Hajarika B J, Pathak K K and Choudhury D K 2011 *Mod. Phys.Lett.* A **26** 1547
[4] Fernandez Francisco M 2000 *Introduction to Perturbation theory in Quantum mechanics* (New York Washington:CRC Press)
[5] Aitchison I J R and Dudek J J 2002 *Eur. J. Phys.* **23** 605
[6] Choudhury D K and Bordoloi N S 2009 *Mod. Phys. Lett.* A **24** 443
[7] Pathak K K and Choudhury D K 2011 *Chinese Physics Lett.* **28** 101201
[8] Eichten E *et al.*, 1975 *Phys. Rev. Lett.* **34** 369
[9] Hazarika B J and Choudhury D K 2010 *Pramana J. Physics* **75** 423
[10] Ebert D, Faustov R N and Galkin V O 2009 *Phys. Rev.* D **79** 114029

Non-singlet structure function $F_2^{NS}(x,t)$ in DGLAP approach

Neelakshi N K Borah [1], D K Choudhury [1,2], and P K Sahariah [3]

[1]Department of Physics, Gauhati University, Guwahati 781014, India.
[2]Physics Academy of the North East (PANE), Physics Department, Gauhati University, Guwahati 781014, India.
[3] Department of Physics, Cotton College, Guwahati 781001, India.

E-mail: nishi_indr@yahoo.co.in

Abstract. In this work the non-singlet structure function $F_2^{NS}(x,t)$ have been obtained by solving DGLAP evolution equation in leading order (LO) at the small x limit. Here we have used a Taylor series expansion and then the method of characteristics and Lagrange's method to solve the evolution equations. We make a detailed comparison of the predictions of the two methods with two different levels of approximations with experimental data as well as with numerical solutions.

1. Introduction

The structure functions of the nucleon are not calculable in QCD but their evolution in Q^2 are predicted by a set of integro-differential equations known as DGLAP equations [1]. Apart from the numerical solution [2], there is the alternative approach of studying analytically these equations at small x and there are many analytical solutions available in literature [3, 4] and the present authors have also pursued such an approach with reasonable phenomenological success [5–7]. In this paper we study analytical solutions of the non-singlet structure functions. We convert the LO DGLAP equation into a partial differential equations in the two variables (x, Q^2) by a Taylor series expansion, with two different levels of approximations, valid at low x. The resulting partial differential equations are then solved analytically by two different methods: Lagrange's auxiliary method [8] and the method of characteristics [9, 10]. The aim of the paper is to make a detailed comparison of the predictions of the two methods, with two different levels of approximations. In section 2 we give the formalism, section 3 is devoted to discussions of the solutions and in section 4 we give our conclusion.

2. Formalism

The evolution equation for non-singlet flavour dependent contribution, which evolve independently in the DGLAP approach can be written as [1],

$$\frac{\delta F_2^{NS}(x,t)}{\delta t} = \frac{A_f}{t}\left[\{3+4\ln(1-x)\}F_2^{NS}(x,t)\right.$$
$$\left.+2\int_x^1 \frac{dz}{1-z}\left\{(1+z^2)F_2^{NS}\left(\frac{x}{z},t\right)-2F_2^{NS}(x,t)\right\}\right]. \quad (1)$$

Here $A_f = 4/(3\beta_0)$ and $\beta_0 = 11 - 2n_f/3$ is QCD beta function at LO. At small x, approximating $F_2^{NS}(\frac{x}{z}, t)$ on RHS of Eq.(1) and defining $u = 1 - z$ [6, 7] we get,

$$\frac{\delta F_2^{NS}(x,t)}{\delta t} = \frac{A_f}{t}\left[\{3 + 4\ln(1-x)\} F_2^{NS}(x,t) + 2\int_x^1 \frac{dz}{1-z}(z^2-1)F_2^{NS}(x,t) + 2\int_x^1 \frac{dz}{1-z}(1+z^2)(x\sum_{k=1}^{\infty} u^k)\frac{\delta F_2^{NS}(x,t)}{\delta x}\right]. \tag{2}$$

Carrying out the integration in z, we can write Eq.(2) as,

$$\frac{\delta F_2^{NS}(x,t)}{\delta t} - \frac{A_f x}{t}\frac{\delta F_2^{NS}(x,t)}{\delta x}\left[2\ln(\frac{1}{x}) + (1-x^2)\right] = \frac{A_f}{t}\left[3 + 4\ln(1-x) + (x-1)(x+3)\right] F_2^{NS}(x,t). \tag{3}$$

Eq.(3) is a partial differential equation for the non-singlet structure function $F_2^{NS}(x,t)$ with respect to the variables x and t.

While performing the integration in z, neglecting terms $\mathcal{O}(x^2)$ and higher, we can also express Eq.(2) as,

$$\frac{\delta F_2^{NS}(x,t)}{\delta t} - \frac{8A_f}{3}\frac{x}{t}\frac{\delta F_2^{NS}(x,t)}{\delta x} = \frac{A_f\{4\ln(1-x)+2x\}}{t} F_2^{NS}(x,t) \tag{4}$$

This we get by considering,

$$x\sum_{k=1}^{\infty} u^k = xu = x(1-z) \tag{5}$$

during integration.

We solve both the PDE Eq.(3) and Eq.(4) with the two formalisms, the Lagrange's method and method of characteristics.Though both these PDE are obtained from the same Eq.(1).

2.1. Solution by the method of characteristics

By adopting the method of characteristics [9, 10], one can express Eq.(3), as an ordinary derivative with respect to t and the equation becomes an ordinary differential equation:

$$\frac{dF_2^{NS}(x(t),t)}{dt} = c^{NS}(x(t),t)\, F_2^{NS}(x(t),t) \tag{6}$$

where

$$c^{NS}(x(t),t) = \frac{A_f\{4\ln(1-x(t)) + (x-1)(x+3)\}}{t} \tag{7}$$

Integrating Eq.(6) over t from t_0 to t along the characteristic curve, one gets the solution for the non-singlet as:

$$F_2^{NS}(x,t) = F_2^{NS}(\tau)\left(\frac{t}{t_0}\right)^{\alpha}, \alpha = \frac{8}{3\beta_0}\{2\ln x - x\} \tag{8}$$

Eq.(8) is the analytical solution of the Eq.(3) within the present formalism. Using the same formalism for the PDE Eq.(4), in a similar way we get a different form of solution for the non-singlet structure function F_2^{NS} as,

$$F_2^{NS}(x,t) = F_2^{NS}(\tau)\left(\frac{t}{t_0}\right)^{\beta}, \beta = \frac{1}{\ln\frac{t}{t_0}}\left[\frac{3}{4A_f}x\left(\left(\frac{t}{t_0}\right)^{\frac{8A_f}{3}} - 1\right) - \frac{3}{2A_f}\left\{\sum_{k=1}^{\infty}\frac{x^k}{k^2}\left(\left(\frac{t}{t_0}\right)^{\frac{8A_f k}{3}} - 1\right)\right\}\right] \tag{9}$$

Eq.(8) and Eq.(9) are the two analytical solutions of Eq.(3) and Eq.(4) respectively.

2.2. Solution by the Lagrange's auxiliary method

To solve the equation Eq.(3) and Eq.(4) by the Lagrange's Auxiliary Method [8], we can write the Eq.(3) and Eq.(4) in the form:

$$Q(x,t)\frac{\delta F_2^{NS}(x,t)}{\delta t} + P(x,t)\frac{\delta F_2^{NS}(x,t)}{\delta x} = R(x,t,F_2^{NS}) \tag{10}$$

The general solution of the Eq.(10) is obtained by solving the following auxiliary system of ordinary differential equations,

$$\frac{dx}{P(x)} = \frac{dt}{Q(t)} = \frac{dF_2^{NS}(x,t)}{R(x,t,F_2^{NS}(x,t))} \tag{11}$$

If $u(x,t,F_2^{NS}) = C_1$ and $v(x,t,F_2^{NS}) = C_2$ are the two independent solutions of Eq.(15), then in general, the solution of Eq.(10) is,

$$F(u,v) = 0 \tag{12}$$

where F is an arbitrary function of u and v.

In this approach we try to find a specific solution that satisfies some physical conditions on the structure function. Such a solution can be extracted from the combination of u and v linear in F_2^{NS},

$$u + \alpha v = \beta \tag{13}$$

where α and β are two quantities to be determined from the boundary conditions on F_2^{NS}. Using the physically plausible boundary conditions, the solution of Eq.(10) takes the form,

$$F_2^{NS}(x,t) = F_2^{NS}(x,t_0)\left(\frac{t}{t_0}\right) \tag{14}$$

In a similar way for the Eq.(4), the Lagrange's method leads us to a solution for the non-singlet structure function F_2^{NS} as given below,

$$F_2^{NS}(x,t) = F_2^{NS}(x,t_0)\left(\frac{t}{t_0}\right)h(x,t) \tag{15}$$

where

$$h(x,t) = \frac{X^{NS}(1) - X^{NS}(x)]}{[(\frac{t}{t_0})X^{NS}(1) - X^{NS}(x)]} \tag{16}$$

with $h(x,t) \leq 1$ for $t \geq t_0$. Here $h(x,t)$ measures deviation of Eq.(15) from solution Eq.(14). We note that the apparent absence of $\log x$ dependence in the solution Eq.(15) is due to algebraic cancellation. Eq.(14) and Eq.(15) are the solutions of the Eq.(3) and Eq.(4) respectively.

3. Results and discussion

We test the validity and compatibility of the two sets of analytical solutions, by comparing them directly with the available data [11] and with the MSTW 08 numerical solutions [12], using the MSTW2008 [13]input. We plot our set of solutions Eq.(8), Eq.(14) and Eq.(9), Eq.(15) respectively with the BCDMS data in Fig.1, where we explore a relatively high x and Q^2 range, $(0.07 \leq x \leq 0.75)$ and $(13\,\mathrm{GeV}^2 \leq Q^2 \leq 63\,\mathrm{GeV}^2)$. From that we observe that though our both set of solutions predict the same behaviour at low value of x and agree with the numerical solutions towards low x range, as we approach the high x range our solutions overshoot both data and the numerical solutions. In case of the solutions obtained by Lagrange's method given by Eq.(14) and Eq.(15), while the Eq.(14) shows logarithmic growth with the increasing Q^2 values, the other solution Eq.(15) remains almost constant with increasing Q^2 for fixed x values. The solutions by Method of Characteristics Eq.(8) and Eq.(9), also show very slow growth with increasing Q^2 for fixed x values.

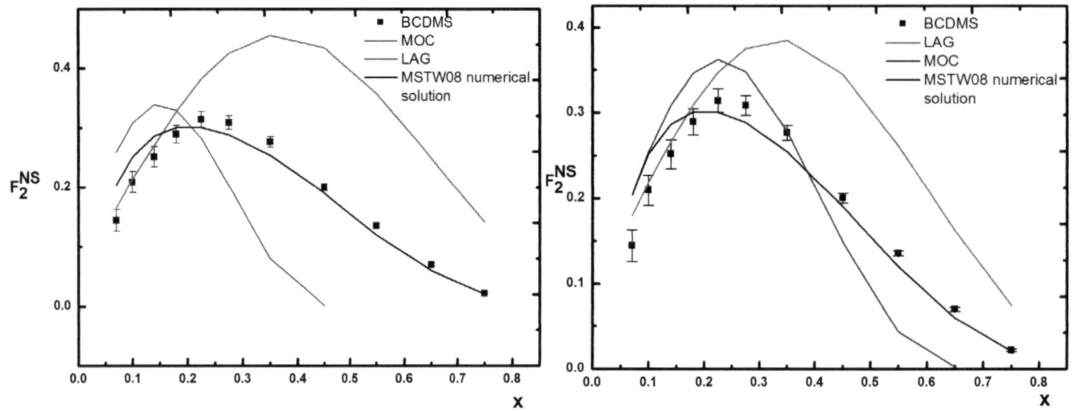

Figure 1. Non-singlet structure function $F_2^{NS}(x,Q^2)$ as function of x at different Q^2 values according to Eq.(8) and Eq.(14) (left) and values according to Eq.(9) and Eq.(15) (right). Data from refs [11].

4. Conclusion

The Taylor approximated DGLAP equation for the non-singlet structure function is solved analytically by two different methods: the Lagrange's auxiliary method and the Method of Characteristics. The quantitative and qualitative differences of the solutions are then discussed. Considering the solutions together, they are valid towards low x region. This demonstrated that two powerful methods of solving differential equations can be successfully applied in the DGLAP framework to obtain analytical solutions. Results of these methods to the polarised structure function $g_1^{NS}(x,t)$ has been reported elsewhere [14].

References

[1] V. N. Gribov and L. N. Lipatov, *Sov. J. Nucl. Phys.* **15** (1972) 438;
L. N. Lipatov, *Sov. J. Nucl. Phys.* **20** (1975) 94;
Yu. L. Dokshitzer, *Sov. Phys. JETP* **46** (1977) 641; G. Altarelli and G. Parisi, *Nucl. Phys.* **B126** (1977) 298

[2] S. Forte, L. Garrido, J. I. Latorre and A. Piccione, *J. H. E. P.* **05** 062 (2002)(hep-ph/0204232;
L Del Debbio, S. Forte, J. I. Latorre, A. Piccione and J. Rojo, *J. H. E. P.* **03** 080 (2005), (hep-ph/0501067)

[3] R. D. Ball and S. Forte, *Phys. Lett.*, **B335** (1994) 77;*ibid* *Phys. Lett.* **B336** (1994) 77

[4] A. V. Kotikov and G. Parente, *Nucl. Phys.* **B549** (1999) 242

[5] D. K. Choudhury and P. K. Sahariah, *Pramana-J.Phys.* **58** (2002) 599;*ibid*

[6] D. K. Choudhury and P. K. Dhar, *Indian J. Phys.* **81** (2) 259 (2007)

[7] D. K. Choudhury and P. K. Dhar, *Indian J. Phys.* **83** (12) 1699-1709 (2009)

[8] I. Sneddon, *Elements of Partial Differential Equations*, M.Graw Hill, NewYork (1957)p131.

[9] W. E. Williams, *Partial Differential Equations*, Clarendon Press,Oxford (1980)

[10] E. C. Zachmanoglou and D. W. Thoe, *Introduction to Partial Differential Equations with Applications*, Williams and Wilkins (1976) L. F. Abbot, W. B. Wood and R. M. Barnett, *Phys Rev.* **D22**(1980)582 *Zeit.Phys.*, **C28** (1991)321

[11] BCDMS Collab: Benvenuti *et. al*, *Phys. Lett.*, **237B** (1990)599 A. D. Martin, R. G. Roberts, W. S. Stirling and R. S. Thorne. *hep-ph/0411040*

[12] The Durham HEP Data Project*http://hepdata.cedar.ac.uk/pdf/pdf3.html*

[13] A. D. Martin, W. S. Stirling, R. S. Thorne and G Watt. *hep-ph/0901.0002v3*

[14] N. N. K. Borah, D. K. Choudhury and P. K. Sahariah, *Non-singlet spin structure function $g_1^{NS}(x,t)$ in the DGLAP approach.* *Pramana-J.Phys.***79** (4) 833-837 (2012)

Sea-quark and gluonic contributions to strange baryons and its properties

M Batra and A Upadhyay

School of Physics and Material Science,Thapar University, Patiala.

E-mail: mbatra310@gmail.com,alka@thapar.edu

Abstract. We treat the hadrons as an ensemble of quark-gluon Fock states where contributions from sea-quarks and gluons can be studied in detail for the properties of low lying baryons. Statistical model is applied to calculate individual probabilities from various scalar, vector and tensor sea components in flavour, spin and colour subspaces for each quark-gluon Fock state. The mass of strange quark is imposed in terms of constraints to the availability of strange quarks in the sea and free energy of gluons in conformity with experimental indications. We calculate multiplicities for different sets of Fock states to compute the role of strange sea for cascade doublet Ξ^+ and Ξ^0. The low energy properties like magnetic moment, weak decay matrix elements and axial coupling constant ratios have been studied. The incorporation of strange quark gluon sea is discussed and they are found to affect the results by almost 46%.The results provide deeper understanding for baryon structure thereby motivating experiments for further inspection, especially the spin distribution among the quarks and gluons.

1. Introduction

In the low energy limit,QCD has a challenging behaviour. The non-perturbative and relativistic nature of quarks and gluons makes the hadronic structures difficult to understand. The non-perturbative mechanism can be made to understand via several models. Although the constituent quark model tends to explain the static properties of a baryonic system, still there is a need of some new degrees of freedom. The extension of valence quarks to sea with quark-antiquark pair is a remarkable step in hadron spectroscopy. A variety of approaches have been developed to understand the behaviour of baryons with three valence quarks and sea consisting of gluons and quark-antiquark pair. These models help to make a deeper understanding to the various properties like weak decay coupling ratios, magnetic moments, masses of hadrons and flavour asymmetry. Latest studies, dedicated towards the study of quark-antiquark pairs and gluon condensates includes statistical balance model [1], lattice QCD [3], effective QCD approaches [2] and meson-cloud based approaches [4–6]. The static properties of hadrons had also been studied through a large variety of experiments [7–9]. Recent experimental studies show that strange quark also make a significant contribution to nucleonic spin which can be studied by using polarised deep-inelastic scattering experiments of electrons or muons from nucleons. The aim of present work is to analyse the probabilities of various properties are analysed by determining the multiplicities in colour, flavour and spin states. The multiplicities are used to obtain the probabilities for each quark-gluon Fock states in the ground state baryons. In section II, we briefly present the theoretical frame. Numerical results are given in section III, putting forward predictions for the probabilities of all Fock states for Cascade baryonic system.

Journal of Physics: Conference Series **481** (2014) 012024 doi:10.1088/1742-6596/481/1/012024

2. Theory

A wave-function ($\psi = \Phi[\mid \phi\rangle \mid \psi\rangle \mid \chi\rangle \mid \xi\rangle)$ including 3-quark wave-function and suitable sea contents is framed keeping in mind the total anti-symmetry of the wave-function where, $\mid \phi\rangle, \mid \psi\rangle, \mid \chi\rangle, \mid \xi\rangle$ is for flavour, spin, colour, and space-time wave-function respectively [13]. Let $\Phi_1^{1/2}$ is the standard SU(3) q^3 wave-function transforming as 56 of SU(6) wave-function and $H_{0,1,2}$ and $G_{0,1,2}$ denotes the spin and colour possibilities of sea with spin and colour combinations of $0, 1, 2$ and $1, 8, 10$ respectively. The total flavour-spin-colour wave function of a spin up baryon which consists of three-valence quarks and sea components can be written as given in [13]. $\mid \phi_1^{\uparrow}\rangle = \frac{1}{N}[\phi_1^{(\frac{1}{2})\uparrow} H_0 G_1 + a_8 \phi_8^{(\frac{1}{2})\uparrow} H_0 G_8 + a_{10} \phi_{10}^{(\frac{1}{2})\uparrow} H_0 G_{\overline{10}} + b_1 [\phi_1^{\frac{1}{2}} \otimes H_1]^{\uparrow} G_1 + b_8 (\phi_8^{\frac{1}{2}} \otimes H_1)^{\uparrow} G_{\overline{10}} + c_8 (\phi_8^{\frac{3}{2}} \otimes H_1)^{\uparrow} G_8 + d_8 (\phi_8^{\frac{3}{2}} \otimes H_2)^{\uparrow} G_8]$ where $N^2 = 1 + a_8^2 + a_{10}^2 + b_1^2 + b_8^2 + b_{10}^2 + c_8^2 + d_8^2$. All the terms have to be written properly with appropriate CG coefficients and by taking into account the symmetry property of the component wave function. To obtain all the coefficients in the wave-function, a general parameterisation method is employed to calculate the different observables in the form of eigen values for baryonic system. Assuming the expansion of hadrons in terms of quark-gluon Fock states where Fock states avail the presence of quark-antiquark pairs multi-connected to gluons. Mathematically, it can be expressed as:

$$\mid h\rangle = \sum_{i,j,k} c_{i,j,k} \mid \{q\}, \{i, j, l\}, \{k\}\rangle, \tag{1}$$

where $\{q\}$ represents the valence quarks of the baryon, i, j, l represent the number of $q\bar{q}(u\bar{u}, d\bar{d}, s\bar{s}$ and k is the number of gluons. The principle states that the transition between any two processes either via splitting or recombination involving any two kind of partons should balance each other. The splitting and recombination procedure involves three kinds of processes and their transition probabilities involve number of gluons and quark-antiquark pairs [10–12].

(i) When $q \rightleftharpoons qg$ is considered. The general expression of probability for these sub-processes can be described as

$$\frac{\rho_{i,j,l,k}}{\rho_{i,j,l,k-1}} = \frac{1}{k} \tag{2}$$

(ii) When both the processes $q \leftrightharpoons qg$ and $g \leftrightharpoons gg$ are included. We can write

$$\frac{\rho_{i,j,l,k}}{\rho_{i,j,l,k-1}} = \frac{3 + 2i + 2j + 2l + k - 1}{(3 + 2i + 2j + 2l)k + \frac{k(k-1)}{2}} \tag{3}$$

(iii)When the processes $g \rightleftharpoons q\bar{q}$ are involved: The transition probabilities involving $g \rightleftharpoons q\bar{q}$ depend upon the valence quark content and differ in all baryons.

The generation of $s\bar{s}$ pair from gluons is restricted due to non-negligible mass of s-quark. The constraint on the free energy of gluon comes in the form of factor $k(1-C_l)^{n-1}$ [12] where k is the number of gluons and n represents the total number of partons present in that state. Therefore, the splitting and recombination for the processes involving $g \rightleftharpoons q\bar{q}$ undergoes SU(3) symmetric breaking in sea where q is for some heavier quark flavour. In general,$n = 3 + 2i + 2j + 2l + k$ and $C_{l-1} = \frac{2M_s}{M_B - sM_s - 2(l-1)M_s}$, M_s is the mass of s-quark and M_B is the mass of the baryon.

$$\frac{\rho_{i,j,l,k}}{\rho_{i,j,l+k,0}} = \frac{(k)(k-1) - -1(1 - C_0)^{n+k-2}}{(l+1)(l+2) - - - (l+k)(l+k+1)} \tag{4}$$

The above expression produces all the probabilities in the form of ratios. All the probabilities in flavour space can be obtained from the table given below for individual particle. Complete expressions for cascade doublet can be written as:

Table 1. Probability Expressions

Baryon	Expression
$\Xi^0(uss)$	$\frac{\rho_{i,j,l+k,0}}{\rho_{0,0,0,0}} = \frac{2}{(i)!(i+1)!(j)!(j)!(l+k)!(l+k+2)!}$
$\Xi^-(dss)$	$\frac{\rho_{i,j,l+k}}{\rho_{0,0,0,0}} = \frac{2}{(i)!(i)!(j)!(j+1)!(l+k)!(l+k+2)!}$

These probabilities in flavour space is further needed to calculate probabilities in spin and colour space [14, 15]. The decomposition of baryonic states in various quark-gluon Fock states with relevant operators acting on the valence and sea part is used to find probabilities for all possible spin and colour sub-states so as to produce spin $\frac{1}{2}$ and colour singlet state of the baryon. The procedure can be applied to states like $\mid \overline{u}ug\rangle, \mid \overline{d}dg\rangle, \mid \overline{s}sg\rangle, \mid \overline{u}u\overline{d}d\rangle, \mid gg\rangle$ etc. More favourable situation occurs when the higher multiplicities are suppressed with the fact that the larger multiplicity leads to higher possibility of interaction and less survival probability. Moreover, non-zero mass of s-quark limits the free energy of gluon and hence states with strange quark-antiquark pairs are assumed to be less probable. The suppression of states with higher multiplicities can be made to archive on phenomenological grounds by assuming the probability of a system being in a particular spin and colour state is inversely proportional to multiplicity of the spin and colour of that state. The probability in terms of multiplicity for light and strange quarks are mentioned in tables 2 and 3.The sum of all probabilities in individual cases provide a suitable normalisation constant and the coefficients in the wave-function for baryon. Static properties of ground state strange baryon octets can be determined from these coefficients.

3. Results and discussions

We aim to choose one set of the strange baryon to compute probabilities of several Fock states in spin, flavour and colour space and to check the validity of the statistical model to study their low energy properties. The study has been divided into three cases to study the dominant role of strange sea with proper assumptions. Case I comes with the assumption that gluon has a possibility to create $q\overline{q}$ pair (including strange quark) which carries the quantum numbers of gluon, hence it suggests the suppressed multiplicity of $q\overline{q}$ and g to have spin 1. In case II we rule out the assumption for pair to have the same quantum numbers as that of gluon and it can be in spin state $0, 1$, or 2. Third case is similar to the first case where we modify the detailed balance principle by ruling out the sub-processes $g \rightleftharpoons q\overline{q}$. Fock states in all three cases are mentioned in the table given below. The table 2 differentiates the impact of presence of strange quark-anti-quark pair in sea in each case. The table 3 shows the probability contribution to sea in terms of different number of strange quark and gluon Fock states. Case I, II and III all shows the lower likelihood for the occupancy of Fock states where $s\overline{s}$ pair is present than those without $s\overline{s}$. In table 2, the probability enhances in all cases by almost 0 to 85% as we exclude strange quark-antiquark pairs. In table 3, where we already have $s\overline{s}$ pair in addition to gluons, the relative probability for a single $s\overline{s}$ pair is large as compared to other combinations and it is least for $\mid s\overline{s}$ possibly with a reason where a larger multiplicity leads to the reduced possibility. The results are shown in table 4, where magnetic moment and weak decay couplings are given on the basis of different assumption. Case I with strange sea favours the experimental values, there by suggesting the gluons in the sea to be generating strange quark condensates with similar quantum numbers. The data shifts by 7 to 46 % approaching the experimental value favouring the strangeness in the sea. Here vector strange sea dominates and flavours the results. Present framework suggests a stronger base to choose the model with suggested cases to verify the experimental and other theoretical values and hence provide a deeper understanding to the strange baryon structure.

Table 2. Fock States with Probabilities [with (a) and without (b) $s\bar{s}$]

Fock	Case-I		Case-II		Case-III	
State	(a)	(b)	(a)	(b)	(a)	(b)
-0.012	0.029	0.165	0.029	0.165	0.111	0.103
-0.027	0.002	0.083	0.003	0.083	0.086	0.059
-0.049	0.009	0.02	0.006	0.03	0.021	0.015
-0.080	0.165	0.165	0.148	0.148	0.135	0.184
-0.080	0.082	0.082	0.073	0.073	0.068	0.092
-0.080	0.023	0.072	0.023	0.072	0.149	0.203
-0.080	0.012	0.036	0.011	0.036	0.075	0.102
-0.080	-0.225	-0.044	0.014	-0.115	-0.073	-0.052
-0.080	-0.225	-0.044	0.014	-0.115	-0.073	-0.052

Table 3. Strange Sea Fock State with Probabilities

Fock State	Case-I	Case-II	Case-III
$s\bar{s}$	0.055	0.049	0.033
$s\bar{s}g$	0.002	0.002	0.050
$s\bar{s}gg$	0.042	0.028	0.027

Table 4. Static Properties at Low Energy [with (a) and without (b) $s\bar{s}$]

Property	Case-I		Case-II		Case-III		Expt. Value [16]
	(a)	(b)	(a)	(b)	(a)	(b)	
μ_{Ξ^-}	-0.51	-0.35	-0.48	-0.48	-0.46	-0.46	-0.65 ± 0.025
$\frac{g_A}{g_V}(\Xi^- \rightarrow \Sigma^0)$	0.89	0.80	0.82	0.82	0.79	0.72	1.29 [17]
$\frac{g_A}{g_V}(\Xi^0 \rightarrow \Lambda^0)$	-0.25	0.19	0.24	0.24	0.23	0.21	0.25 ± 0.5
$\frac{g_A}{g_V}(\Xi^- \rightarrow \Xi^0)$	-0.25	-0.27	-0.22	-0.22	-0.21	-0.19	-0.31 [18]

References

[1] Zhang B and Zhang Y J 2010 *Phys. Rev.* D **82**074021
[2] Young R D and Thomas A W 2010*Nucl. Phys.*B **844**266C
[3] Bali G S et al.(QCDSF Collaboration), 2012 *Prog. Part. Nucl. Phys.* **67**467
[4] Bukardt M and Warr B 1992*Phys. Rev. D* **45** 958
[5] Brodsky S J and Ma B Q 1998*Phys. Lett.***381**317
[6] Navarra F S,Nielsen M and Paiva S 1997 *Phys. Rev.* D **56**3041
[7] Arnedo M et al.[EMC] 1989 *Nucl. Phys.* B **321** 541
[8] Nardo L D et al.[HERMES Collaboration]2007 *AIP Proc.* **915**507
[9] Airapetian A et al.[HERMES Collaboration]1989 *Phys. Rev.* D **75**012007
[10] Zhang Y J, Deng W Z and Ma B Q 2002 *Phys. Lett.* B **523**260
[11] Zhang Y J,Ma B Q,Yang L M 2003*Int. J. Mod. Phys.* A **8** 1465
[12] Zhang Y J,Zhang B and Ma B Q 2002 *Phys. Rev.* D **65** 114005
[13] Song X and Gupta V 1994 *Phys. Rev.* D **49** 2211
[14] Singh J P and Upadhyay A 2004 *J. Phys.* G **30** 881
[15] Batra M and Upadhyay A 2012 *Nucl. Phys.* A **18** 2012
[16] Nakumara K et al.,2010 *J. Phys.* G **37** 075021
[17] Kim H,Praszalowicz M,Goeke K 1998 *Phys. Rev.* D **57**299
[18] Kim H,Praszalowicz M,Goeke K 2001 *Acta Physica Polonica* B **32**1343

Self-similarity based model of double parton distribution functions at LHC

Akbari Jahan and D K Choudhury

Department of Physics, Gauhati University, Guwahati 781014, India.

E-mail: akbari.jahan@gmail.com

Abstract. The notion of self-similarity, pursued earlier for small x physics at HERA, is used to construct a model for double parton distribution functions (dPDFs) at small x_1 and x_2, the longitudinal fractional momenta of two partons. A necessary input to such a theoretical approach is the parton distribution functions (PDFs). In this model, the most general form of dPDF is found to have total thirteen parameters to be fitted from LHC data. We show that the constructed dPDF does not factorize into two single PDFs in conformity with QCD expectation and it vanishes at the kinematic boundary.

1. Introduction

Double Parton Distribution Functions (dPDFs) [1–5] are the simplest distribution functions that occur in the multi-partonic interactions (MPI) and are of great relevance for LHC physics as they represent a background for the search of new physics. They are equally important in the study of double parton scattering cross-sections. In this paper, we use the notion of self-similarity based model of proton structure function $F_2\left(x, Q^2\right)$ [6] to construct a Transverse Momentum Dependent double Parton Density Function (TMD dPDF) and a dPDF at small x_1 and x_2, the longitudinal fractional momenta of two partons. Parton distribution functions (PDFs) [7] play the role of a necessary input to such a theoretical approach. They are among the most important sources of information on hadron structure at the level of quarks and gluons. We also investigate if the constructed dPDF has a self-similar behavior at small x_1 and x_2 while it vanishes at the kinematic boundary $x_1 + x_2 = 1$.

2. Formalism

2.1. *Self-similarity based TMD dPDF and dPDF at small x_1 , x_2*

The self-similarity based model of the nucleon structure function proposed in Ref. [6] had been designed to be valid in the kinematical region $6.2 \times 10^{-7} \leq x \leq 10^{-2}$ and $0.045 \leq Q^2 \leq 120\,\mathrm{GeV}^2$. It, however, did not take into account the large x behavior of the PDF [8] which is not unexpected. Till date, there is no phenomenological justification of self-similarity at large x. But it is not unreasonable to assume that self-similarity does not terminate abruptly at $x \approx 0.01$, but smoothly vanishes at $x = 1$, the valence quark limit of proton with no trace of self-similarity at all. We consider this point of view in our present work.

Unlike in Ref. [6], where there is only one hard scale k_t^2 in the TMD PDF, TMD dPDF has two hard scales $k_t^{2(1)}$ and $k_t^{2(2)}$ of partons carrying fractional momenta x_1 and x_2 of flavors i

and j respectively. Corresponding to the virtuality Q^2 of deep inelastic scattering (DIS), the partons also have hard scales $Q^{2(1)}$ and $Q^{2(2)}$. For simplicity, we assume them to be fixed. Following Ref. [6], the TMD dPDF for partons of flavors i and j will then have the following basic magnification factors:

$$\begin{aligned} M_1 &= \frac{1}{x_1} \\ M_2 &= \frac{1}{x_2} \\ M_3 &= \left(1 + \frac{k_t^{2(1)}}{k_0^2}\right) \\ M_4 &= \left(1 + \frac{k_t^{2(2)}}{k_0^2}\right) \end{aligned} \tag{1}$$

where k_0^2 is the transverse mass cut-off parameter and has been introduced to take care of the dimension.

The TMD dPDF for a pair of partons of flavors i and j is then given as:

$$\begin{aligned} \log f_{ij}\left(x_1, x_2, k_t^{2(1)}, k_t^{2(2)}\right) &= D_1 \log M_1 \log M_2 + D_2 \log M_1 \log M_4 + D_3 \log M_2 \log M_3 \\ &+ D_4 \log M_3 \log M_4 + D_5 \log M_1 \log M_2 \log M_3 \\ &+ D_6 \log M_1 \log M_2 \log M_4 + D_7 \log M_1 \log M_3 \log M_4 \\ &+ D_8 \log M_2 \log M_3 \log M_4 + D_9 \log M_1 \log M_2 \log M_3 \log M_4 \\ &+ D_0^{ij} - \log M^4 \end{aligned} \tag{2}$$

Consequently, the TMD dPDF is integrated over $Q^{2(1)}$ and $Q^{2(2)}$ to obtain the desired dPDF. It is given as

$$D_{ij}\left(x_1, x_2, Q^{2(1)}, Q^{2(2)}\right) = \frac{e^{D_0^{ij}}}{M^4} \left(\frac{1}{x_1}\right)^{D_1 \log \frac{1}{x_2}} I\left(x_1, x_2, Q^{2(1)}, Q^{2(2)}\right) \tag{3}$$

where $I\left(x_1, x_2, Q^{2(1)}, Q^{2(2)}\right)$ is the double integration over $k_t^{2(1)}$ and $k_t^{2(2)}$ and its expression is given by,

$$I\left(x_1, x_2, Q^{2(1)}, Q^{2(2)}\right) = \int_0^{Q^{2(1)}} dk_t^{2(1)} \int_0^{Q^{2(2)}} dk_t^{2(2)} \left(\frac{k_t^{2(1)} + k_0^2}{k_0^2}\right)^A \left(\frac{k_t^{2(2)} + k_0^2}{k_0^2}\right)^B, \tag{4}$$

where

$$\begin{aligned} A &= D_3 \log \frac{1}{x_2} + D_4 \log\left(\frac{k_t^{2(2)} + k_0^2}{k_0^2}\right) + D_5 \log \frac{1}{x_1} \log \frac{1}{x_2} + D_7 \log \frac{1}{x_1} \log\left(\frac{k_t^{2(2)} + k_0^2}{k_0^2}\right) \\ B &= D_2 \log \frac{1}{x_1} + D_6 \log \frac{1}{x_1} \log \frac{1}{x_2} + D_8 \log \frac{1}{x_2} \log\left(\frac{k_t^{2(1)} + k_0^2}{k_0^2}\right) \\ &+ D_9 \log \frac{1}{x_1} \log \frac{1}{x_2} \log\left(\frac{k_t^{2(1)} + k_0^2}{k_0^2}\right) \end{aligned}$$

The self-similar dPDF at small x_1 and x_2 obtained in Eq.(3) therefore contains 12 parameters, viz. 9 fractal parameters ($D_1, ..., D_9$), one normalization constant D_0^{ij}, one mass scale M^4 (which has been introduced in Eq.(2) so as to make the dPDF dimensionless) and one transverse mass cut-off parameter k_0^2. We note that the integration [Eq.(4)] is not factorisable in x_1 and x_2. Thus the usual factorisability assumption [9] that a dPDF can be considered as a product of two single PDFs does not hold good in the present self-similarity based dPDF.

2.2. *Self-similarity based TMD dPDF and dPDF at the boundary $x_1 + x_2 = 1$*

The standard behavior of single PDF is

$$\lim_{x_1 \to 1} D_i \left(x_1, Q^{2(1)}\right) = 0 \,\text{and}\, \lim_{x_2 \to 1} D_j \left(x_2, Q^{2(2)}\right) = 0 \tag{5}$$

The corresponding behavior of dPDF at the kinematic boundary, on the other hand, is [1]

$$\lim_{x_1 + x_2 \to 1} D_{ij} \left(x_1, x_2, Q^{2(1)}, Q^{2(2)}\right) = 0 \tag{6}$$

It is to be noted that Eq.(5) and Eq.(6) do not conform with each other indicating again that the simple assumption of factorisability of dPDF into two PDFs fails at the kinematic boundary $x_1 + x_2 = 1$. Thus the present self-similarity based model of dPDF conforms to the QCD expectation of non-factorisability [10].

As mentioned above, the notion of self-similarity for dPDF at large x is not expected to hold. However, a simple parameter-free way of incorporating the kinematic boundary condition [Eq.(6)] is to introduce an additional magnification factor $M_5 = \left(\frac{1}{x_1+x_2} - 1\right)$, which for $x_2 = 0$ and $x_1 \to 0$ ($x_1 = 0$ and $x_2 \to 0$) approaches $M_1 (M_2)$ of Eq.(1). The corresponding dPDF thus obtained is

$$\tilde{D}_{ij} \left(x_1, x_2, Q^{2(1)}, Q^{2(2)}\right) = \frac{e^{D_0^{ij}}}{M^4} \left(\frac{1}{x_1}\right)^{D_1 \log \frac{1}{x_2}} \left(\frac{1 - (x_1 + x_2)}{x_1 + x_2}\right)^{D_{10}} \tilde{I} \left(x_1, x_2, Q^{2(1)}, Q^{2(2)}\right) \tag{7}$$

where $\tilde{I} \left(x_1, x_2, Q^{2(1)}, Q^{2(2)}\right)$ is the double integration over the transverse momenta $k_t^{2(1)}$ and $k_t^{2(2)}$ and has the same expression as that of $I \left(x_1, x_2, Q^{2(1)}, Q^{2(2)}\right)$ as given in Eq.(4). The final expression of dPDF at the kinematic boundary in this approach thus contains total thirteen parameters (12 parameters as mentioned above and one additional fractal parameter D_{10}) to be determined from LHC data.

3. Graphical analysis of dPDFs

The simplest model of dPDF for small x_1 and x_2 can be obtained if we assume that D_1, $D_{10} \gg D_2, \ldots, D_9$. The dPDF expression of Eq.(3) will then be reduced to

$$D_{ij} \left(x_1, x_2, Q^{2(1)}, Q^{2(2)}\right) \simeq \frac{e^{D_0^{ij}}}{M^4} \left(\frac{1}{x_1}\right)^{D_1 \log \frac{1}{x_2}} Q^{2(1)} Q^{2(2)} \tag{8}$$

The dPDF at the kinematic boundary [Eq.(7)] also gets reduced to

$$\tilde{D}_{ij} \left(x_1, x_2, Q^{2(1)}, Q^{2(2)}\right) \simeq \frac{e^{D_0^{ij}}}{M^4} \left(\frac{1}{x_1}\right)^{D_1 \log \frac{1}{x_2}} \left(\frac{1 - (x_1 + x_2)}{x_1 + x_2}\right)^{D_{10}} Q^{2(1)} Q^{2(2)} \tag{9}$$

For qualitative feature of the model, we plot in Figure 1 the dPDF vs x for a few representative values of $x_2 = 0.3$, 0.1 and 0.6 using Eq.(8) (dashed lines) and Eq.(9) (solid lines). It shows the qualitative difference between the self-similarity based model of Ref.[6] and the smooth extrapolation for large x.

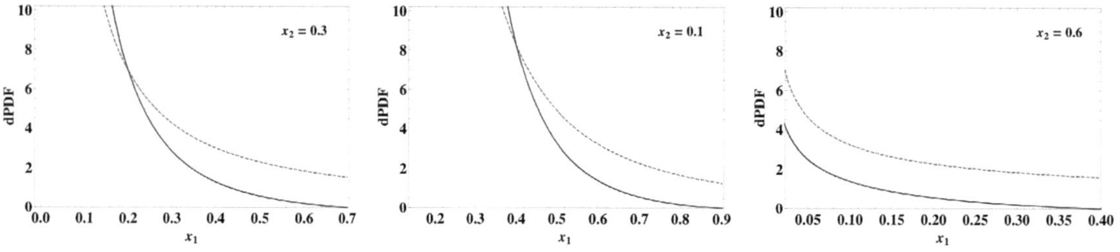

Figure 1. Double Parton Distribution Functions (dPDFs) shown as a function of x.

4. Conclusions

In this paper, we have introduced the self-similarity based formalism of dPDF at small x_1, x_2. We have found that the constructed dPDF does not factorize into two single PDFs in conformity with the QCD expectation. We have also suggested a smooth continuous form of dPDF which has both the expected self-similar behavior at small x_1 and x_2 and vanishes at the kinematic boundary. This is achieved by introducing an additional factor in the defining TMD dPDF.

References

[1] Bartalini P *et al* 2011 Multi-Parton Interactions at the LHC *Preprint* 1111.0469[hep-ph]
[2] Gaunt J R and Stirling W J 2010 *JHEP* **03** 005 (*Preprint* 0910.4347[hep-ph])
[3] Snigirev A M 2010 *Phys. Rev.* D **81** 065014 (*Preprint* 1001.0104[hep-ph])
[4] Snigirev A M 2011 *Phys. Rev.* D **83** 034028 (*Preprint* 1010.4874[hep-ph])
[5] Chang H M, Manohar A V and Waalewijn W J 2013 *Phys. Rev.* D **87** 034009 (*Preprint* 1211.3132[hep-ph])
[6] Lastovicka T 2002 *Euro. Phys. J.* C **24** 529 (*Preprint* hep-ph/0203260)
[7] Roberts R G 1990 *The Structure of the Proton* (Cambridge: Cambridge University Press) p 120
[8] Yndurain F J 1992 *Theory of Quark and Gluon Interactions* (Berlin: Springer) p 129
[9] Bartels J 2011 *Proc. of the 3rd Int. Workshop on Multiple Partonic Interactions at the LHC*, ed S Platzer and M Diehl (DESY) p 151
[10] Snigirev A M 2003 *Phys. Rev.* D **68** 114012 (*Preprint* hep-ph/0304172)

Solution of nonlinear Gribov-Levin-Ryskin-Mueller-Qiu evolution equation for gluon distribution function

Mayuri Devee and J K Sarma

HEP Laboratory, Department of Physics, Tezpur University, Tezpur, Assam, 784 028, India.

E-mail: deveemayuri@gmail.com

Abstract. In this paper we have determined the behavior of gluon distribution function by solving the Gribov-Levin-Reskin-Mueller-Qiu (GLR-MQ) evolution equation,which is nonlinear in gluon density. The moderate Q^2 behavior of $G(x,t)$, where $t = \ln(Q^2/\Lambda^2)$, is obtained by employing the Regge like behaviour of gluon distribution function at small-x. Here Q^2 behavior of nonlinear gluon distribution function is investigated for small values $x = 10^{-2}$, 10^{-3}, 10^{-4} and 10^{-5} rexpectively. Our predictions are compared with different parametrisations and are found in good agreement. It is observed from our results that with the nonlinear corrections incorporated, the strong growth of $G(x,t)$ that corresponds to the linear QCD evolution equation is slowed down. Moreover essential taming of gluon distribution function is observed for $R = 2$ GeV^{-1} as expected.

1. Introduction

Studies of parton distribution function or more importantly the gluon distribution function can provide an exquisitely fine test of perturbative quantum chromodynamics (QCD) in the small-x region and can disclose a systematic examination of new effects such as saturation of parton (quark and gluon) density in the nucleon. The determination of the gluon density at small values of the Bjorken scaling variable x is particularly interesting, because in this region gluons are expected to dominate the proton structure function. The most precise determinations of the gluon momentum distribution in the proton can be obtained from a measurement of the deep inelastic scattering (DIS) proton structure function $F_2(x,t)$ and its scaling violation. In perturbative QCD, the high-Q^2 behavior of DIS has traditionally been determined by the linear Dokshitzer-Gribov-Lipatov-Altarelli-Parisi (DGLAP) evolution equations [1]. Consequently, the approximate analytical solutions of DGLAP evolution equations have been reported in recent years with significant phenomenological success [2,3]. The sharp growth of the gluon distribution function as x goes smaller observed at HERA [4] can ultimately violate unitarity and so it has to be controlled by considering the gluon recombination or shadowing effects at very small values of x. This leads to the modification of the evolution equation by a term which is non-linear in gluon density. Gribov, Levin, Ryskin, Mueller and Qiu first studied these effects and proposed a new nonlinear evolution equation known as the GLR-MQ equation [5,6]. The study of viable generalizations of the GLR-MQ equation has been performed with great interest in the last few years [7-10].

In the present work we provide a solution of the nonlinear GLR-MQ evolution equation for the calculation of $G(x,t)$ in leading order by employing Regge behavior of gluon distribution function. The moderate Q^2-behavior of nonlinear gluon distribution function is obtained and results are compared with the predictions of different paramerisations viz. GRV98 [11], MRST2001 [12] and EHKQS model [7] etc. Finally, we present our conclusions.

2. Solution of nonlinear GLR-MQ evolution equation

The nonlinear GLR-MQ equation is given as [13]

$$\frac{\partial G(x,Q^2)}{\partial \ln Q^2} = \frac{\partial G(x,Q^2)}{\partial \ln Q^2}\bigg|_{\rm DGLAP} - \frac{9}{2}\frac{\alpha_S^2(Q^2)}{R^2Q^2}\int_x^1 \frac{d\omega}{\omega}\Big[G\Big(\frac{x}{\omega},Q^2\Big)\Big]^2. \tag{1}$$

The size of the nonlinear term depends on the value of R where, R is the correlation radius between two interacting gluons. The negative sign in front of the nonlinear term is responsible for gluon recombination.

To solve the GLR-MQ equation we consider a simple form of Regee like behavior of the gluon distribution function as

$$G(x,t) = M(t)x^{-\lambda}, \tag{2}$$

where $M(t)$ depends only on t, where $t = \ln(Q^2/\Lambda^2)$. Now using this behavior and substituting the expression for the usual DGLAP term [14], Eq. (1) can be expressed as

$$\frac{\partial M(t)}{\partial t} = \alpha_S(t)M(t)\Big[\Big(\frac{11}{12} - \frac{N_f}{18} + \ln(1-x)\Big) + \int_x^1 d\omega\Big\{\frac{\omega^{\lambda+1}-1}{1-\omega} +$$
$$\Big(\omega(1-\omega) + \frac{1-\omega}{\omega}\Big)\omega^{\lambda}\Big\}\Big] - \frac{9}{2}\frac{\alpha_S^2(t)}{R^2\Lambda^2 e^t}M^2(t)\int_x^1 \omega^{2\lambda-1}d\omega. \tag{3}$$

Performing the integrations explicitly and rearranging the terms, equation (3) takes the form

$$\frac{\partial M(t)}{\partial t} = P(x)\frac{M(t)}{t} - Q(x)\frac{M^2(t)}{t^2e^t}, \tag{4}$$

which can be solved as

$$M(t) = \frac{t^{P(x)}}{C - Q(x)\Gamma[-1+P(x),t]}. \tag{5}$$

Here,Γ is the incomplete gamma function and C is a constant. $P(x)$ and $Q(x)$ are functions of x. However, the dependence of the functions $P(x)$ and $Q(x)$ on x can be considered as mild which could be neglected in the region of small-x. Thus this solution is valid only for small-x.

Substituting Eq. (5) in Eq. (2) we get

$$G(x,t) = \frac{t^{P(x)}}{C - Q(x)\Gamma[-1+P(x),t]}x^{-\lambda}. \tag{6}$$

Now, defining

$$G(x,t_0) = \frac{t_0^{P(x)}}{C - Q(x)\Gamma[-1+P(x),t_0]}x^{-\lambda}, \tag{7}$$

at $t = t_0$ for some lower value of $Q = Q_0$, we obtain from Eq. (6)

$$G(x,t) = \frac{t^{P(x)}G(x,t_0)}{t_0^{P(x)}x^{-\lambda} + Q(x)\Big\{\Gamma[-1+P(x),t_0] - \Gamma[-1+P(x),t]\Big\}G(x,t_0)}x^{-\lambda}, \tag{8}$$

which gives the Q^2 behavior of nonlinear gluon distribution function at fixed x in leading order.

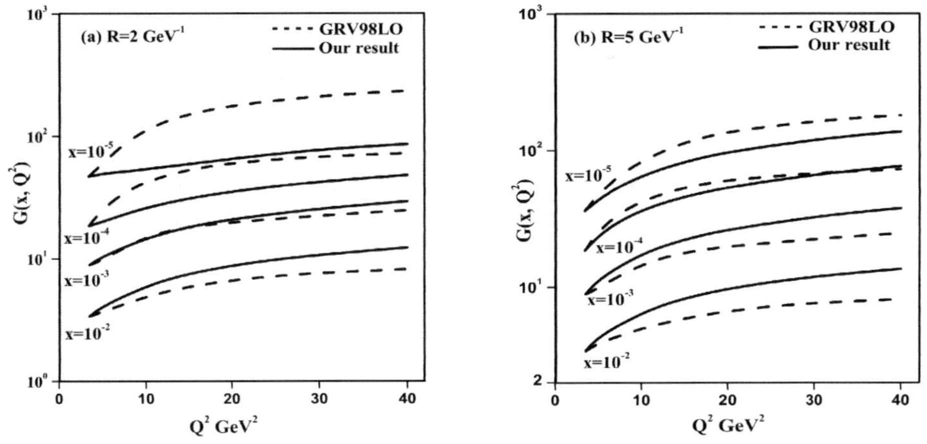

Figure 1. $G(x, Q^2)$ versus Q^2 for $R = 2\ \text{GeV}^{-1}$ and $R = 5\ \text{GeV}^{-1}$ using Eq. (8) for four small values of x compared with GRV98LO (dashed lines).

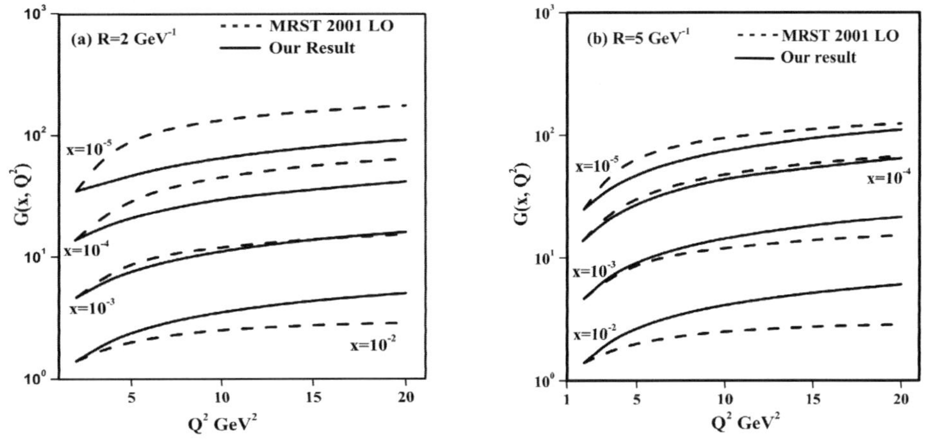

Figure 2. $G(x, Q^2)$ versus Q^2 for $R = 2\ \text{GeV}^{-1}$ and $R = 5\ \text{GeV}^{-1}$ using Eq. (8) for four small values of x compared with MRST2001LO (dashed lines).

3. Result and discussion

Figures 1a,b; 2a,b and 3a,b respectively represents the plot of $G(x, Q^2)$ versus Q^2 for $R = 2\ \text{GeV}^{-1}$ and $R = 5\ \text{GeV}^{-1}$ obtained from Eq. (8) for small values of $x = 10^{-2}$, 10^{-3}, 10^{-4} and 10^{-5}. Our predictions of $G(x, Q^2)$ are compared with those obtained by the global fits to the parton distribution functions GRV98LO, MRST2001LO and EHKQS model. In order to find out the input distribution $G(x, t_0)$, we use MRST2001 LO parton distributions at $Q_0^2 = 1\ \text{GeV}^2$. In our analysis we consider the range $3\ \text{GeV}^2 < Q^2 < 40\ \text{GeV}^2$ for GRV98LO, $1.5\ \text{GeV}^2 < Q^2 < 20\ \text{GeV}^2$ for MRST2001LO and $1.4\ \text{GeV}^2 < Q^2 < 40\ \text{GeV}^2$ for EHKQS.

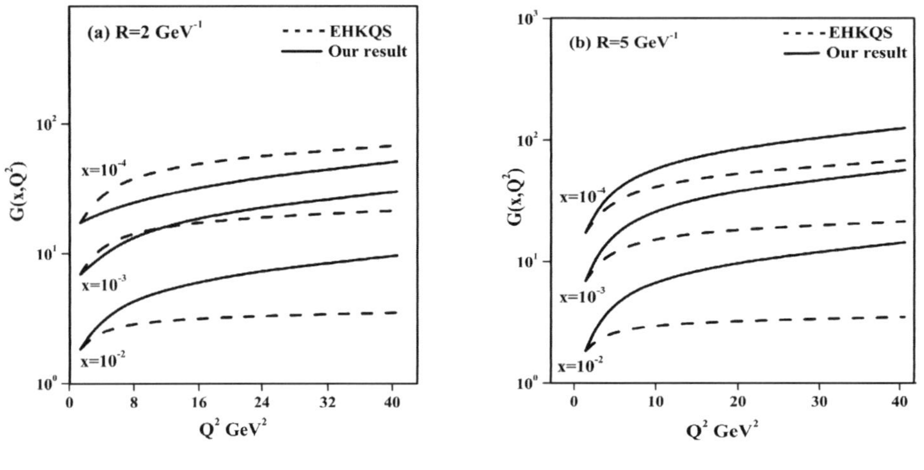

Figure 3. $G(x, Q^2)$ versus Q^2 for $R = 2$ GeV^{-1} and $R = 5$ GeV^{-1} using Eq. (8) for four small values of x compared with EHKQS model (dashed lines).

4. Conclusion

In this paper we have obtained the Q^2 behavior of nonlinear gluon distribution function at fixed x by solving GLR-MQ evolution equation and made an analysis of GRV98LO, MRST2001LO parametrisations and EHKQS model with it. We observe that for fixed-x the gluon distribution function increases with increasing Q^2 as usual, but with the nonlinear corrections incorporated, the strong growth of gluon distribution function is eventually slowed down. Moreover essential taming of gluon distribution function is observed for $R = 2$ GeV^{-1} as expected.

References

[1] Dokshitzer Y L 1977 *Sov. Phys. JETP* **46** 641; Altarelli G and Parisi G 1977 it Nucl. Phys. B **126** 298; Gribov V N and Lipatov L N 1972 *Sov. J. Nucl. Phys.* **15** 438
[2] Botje M 2000 *Eur. Phys. J.* C **14** 285; Vogt A 2005 *Comput. Phys. Commun.* **170** 65; Cafarella A , Coriano C and Guzzi M 2006 *Nucl. Phys.* B **748** 253
[3] Devee M, Baishya R and Sarma J K 2012 *Eur. Phys. J.* C **72** 2036; Baishya R and Sarma J K 2009 *Eur. Phys. J.* C **60** 585; Shah N H and Sarma J K 2008 *Phys. Rev.* D **77** 074023
[4] Adloff C et al (H1 Collaboration) 2001 *Phys. Lett.* B **520** 183
[5] Gribov L V, Levin E M and Ryskin M G 1983 *Phys. Rep.* **100** 1
[6] Mueller A H and Qiu J 1986 *Nucl. Phys.* B **268** 427
[7] Eskola K J, Honkanen H, Kolhinen V J, Qiu J and Salgado C A 2003 *Nucl. Phys.* B **660** 211; Eskola K J, Kolhinen V J and Vogt R 2004 *J. Phys. G: Nucl. Part. Phys.* **30** S1171
[8] Prytz K 2001 *Eur. Phys. J.* C **22** 317; Prytz K 2003 *J. Phys. G: Nucl. Part. Phys.* **29** 285
[9] Mariotto C B and Goncalves V P 2007 *Int. J. Mod. Phys.* E **16** 2984
[10] Rezaei B and Boroun G R 2010 *Phys. Lett.* B **692** 247
[11] Gluck M, Reya E and Vogt A Z. Phys. 1998 *Eur. Phys. J.* C **5** 461
[12] Martin A D, Roberts R G, Stirling W J and Thorne R S 2002 *Phys. Lett.* B **531** 216
[13] Laenen E and Levin E 1995 *Nucl. Phys.* B **451** 207
[14] Abott L F, Atwood W B and Barnett R M 1980 *Pys. Rev. D* **22** 582

Exactly solvable central potentials related to Romanovski polynomials

Nabaratna Bhagawati

Department of Physics, Gauhati University, Guwahati-781014, India

E-mail: nabaratna2008@gmail.com

Abstract. In this paper, we apply a simple transformation method to construct exactly solvable potentials of Schrödinger equation in any arbitrary dimensional Euclidean space. The normalized wave functions of the constructed potentials are obtained in terms of Romanovski polynomials.

1. Introduction

Exact solution of Schrödinger equation with a physical potential is of utmost importance in non-relativistic quantum mechanics. Successful solution of Schrödinger equation provides analytical form of the normalized wave function and quantized energy eigenvalues. However, a very few quantum systems yield exact solutions for potentials of physical interest. Along the years, many authors have tried different methods to obtain the exact solution of the Schrödinger equation [1–6]. To get the bound state wave functions in terms of Romanovski polynomials, we apply a simple transformation method [6] which comprises of a co-ordinate transformation followed by a functional transformation. By applying this method, we transform the second order ordinary differential equation satisfied by Romanovski polynomials to standard radial Schrödinger equation in D-dimensional Euclidean space and thus try to construct exactly solvable potentials. The motivation for doing this work comes from the fact that exact solutions of Schrödinger equation in terms of Romanovski polynomials is not so widespread [7,8] like other orthogonal polynomials [1–5,9,10].

The paper is organized as follows. In section 2, the formalism of the theory is given. In section 3, exact solution of Schrödinger equation in terms of Romanovski polynomials is discussed. The conclusions are discussed in section 4.

2. Formalism

We start with the linear second order differential equation satisfied by a special function $Q(r)$

$$Q''(r) + M(r)Q'(r) + J(r)Q(r) = 0 \tag{1}$$

where a prime denotes differentiation with respect to its argument. $Q(r)$ will later be identified as one of the orthogonal polynomials.

The transformation method comprises of the following two steps

$$r \to g(r) \quad \text{and} \quad \psi(r) = f^{-1}(r)Q(g(r)) \tag{2}$$

We implement the above prescription to equation (1) and obtain

$$\psi''(r) + \left[\frac{d}{dr}\ln\frac{f^2(r)\exp(\int M(g)dg)}{g'(r)}\right]\psi'(r) +$$
$$\left(\frac{f''(r)}{f(r)} - \frac{g''(r)}{g'(r)}\frac{f'(r)}{f(r)} + g'(r)M(g)\frac{f'(r)}{f(r)} + g'^2 J(g)\right)\psi(r) = 0 \tag{3}$$

The radial Schrödinger equation in D-dimensional Euclidean space is($\hbar = 1 = 2m$)

$$\psi''(r) + \frac{(D-1)}{r}\psi'(r) + \left(E_n - V(r) - \frac{\ell(\ell+D-2)}{r^2}\right)\psi(r) = 0 \tag{4}$$

Consistency of equations (3) and (4) demand that

$$\frac{d}{dr}\ln\frac{f^2(r)\exp(\int M(g)dg)}{g'(r)} = \frac{(D-1)}{r} \tag{5}$$

which fixes the form of $f(r)$ as

$$f(r) = Nr^{\frac{(D-1)}{2}} g'^{\frac{1}{2}}\left[\exp\left(-\int M(g)dg\right)\right]^{\frac{1}{2}} \tag{6}$$

where N is the integration constant and plays the role of the normalization constant of the wave functions.
Using (5) and (6) in equation (3) yields

$$\frac{\psi''(r)}{\psi(r)} + \frac{(D-1)}{r}\frac{\psi'(r)}{\psi(r)} = -\frac{1}{2}\{g,r\} + \frac{g'^2(r)}{4}\left[M^2(g) + 2M'(g) - 4J(g)\right] - \frac{(D-1)(D-3)}{4r^2} \tag{7}$$

where the Schwartzian derivative symbol [11], $\{g,r\}$ is defined as

$$\{g,r\} = \frac{g'''(r)}{g'(r)} - \frac{3}{2}\frac{g''^2(r)}{g'^2(r)}$$

From equations (2) and (6), the expression for normalisable wave function is

$$\psi(r) = Nr^{-\frac{(D-1)}{2}} g'^{-\frac{1}{2}}\left[\exp\left(\int M(g)dg\right)\right]^{\frac{1}{2}} Q(g(r)) \tag{8}$$

The radial wave function $\psi(r) = \frac{u(r)}{r}$ has to satisfy the boundary condition $u(r) = 0$, in order to rule out singular solutions [12].

Expression (7) can be cast in the standard Schrödinger equation form (equation (4)) if we can write

$$-(E_n - V(r)) = -\frac{1}{2}\{g,r\} + \frac{g'^2(r)}{4}\left[M^2(g) + 2M'(g) - 4J(g)\right] - \frac{(D-1)(D-3)}{4r^2} \tag{9}$$

Once we choose a particular orthogonal polynomial, $Q(g)$, to construct an exact solution of the Schrödinger equation, the characteristic functions of the polynomial $M(g)$, $J(g)$ get specified. We have to choose one or more than one terms containing the function $g(r)$ in expression (9) and put it equal to a constant to get the energy eigenvalues E_n. The procedure is worked out in detail for Romanovski polynomials in the next section.
It is interesting to note that when the generated potential is purely non-power law, the potential given by expression (9) has a term $(D-1)(D-3)/(4r^2)$ which behaves as constant background attractive inverse square potential in any arbitrary dimension except for dimensions 1 and 3. For power law cases, this background potential and the potential coming from Schwartzian derivative unite to give the correct centrifugal barrier potential in arbitrary dimensions.

3. Construction of exactly solvable potentials from Romanovski polynomials

Identifying

$$Q(g(r)) = R_n^{(\alpha,\beta)}(g) \tag{10}$$

as the Romanovski polynomial and its characteristic functions $M(g)$ and $J(g)$ are

$$M(g) = \frac{2(1-\alpha)g + \beta}{1+g^2} \quad ; \quad J(g) = -\frac{n(n-1) + 2n(1-\alpha)}{1+g^2} \tag{11}$$

Using equations (10) and (11) in equation (7) yields

$$\begin{aligned}\frac{\psi''(r)}{\psi(r)} + \frac{(D-1)}{r}\frac{\psi'(r)}{\psi(r)} &= \frac{1}{4}(4(1-\alpha)+\beta^2)\frac{g'^2}{(1+g^2)^2} + (n(n-1)+2n(1-\alpha))\frac{g'^2}{1+g^2} \\ &- \alpha\beta\frac{gg'^2}{(1+g^2)^2} - \alpha(1-\alpha)\frac{g^2g'^2}{(1+g^2)^2} - \frac{1}{2}\frac{g'''}{g'} + \frac{3}{4}\frac{g''^2}{g'^2} - \frac{(D-1)(D-3)}{4r^2}\end{aligned} \tag{12}$$

and equation (8) yields

$$\psi(r) = Nr^{-\frac{(D-1)}{2}} g'^{-\frac{1}{2}}(1+g^2)^{\frac{1-\alpha}{2}} \exp\left(\frac{\beta}{2}\tan^{-1} g\right) R_n^{(\alpha,\beta)}(g) \tag{13}$$

To convert equation (12) into a standard stationary state Schrödinger equation, we make one or more terms of the right hand side of equation (12) a constant quantity. This enables us to get the energy eigenvalues E_n, the functional form of $g(r)$ and subsequently potential $V(r)$ and wave function $\psi(r)$.

(i) As a first case, let us choose

$$\frac{g'^2}{(1+g^2)^2} = p_1^2 \tag{14}$$

where p_1^2 is a real positive constant independent of r. Equation (14) gives the functional form of $g(r)$ as $g(r) = \tan p_1 r$. Using this value in equation (12) yields

$$E_n = -\left[n(n-1) - \alpha(2n-1) - 2\alpha - 1 + \frac{\beta^2}{4}\right] p_1^2 \tag{15}$$

$$V(r) = Ap_1^2 \tan^2 p_1 r + Bp_1^2 \tan p_1 r - \frac{(D-1)(D-3)}{4r^2} \tag{16}$$

$$\psi(r) = Nr^{-\frac{(D-1)}{2}} (\sec p_1 r)^{-(\alpha+1)} \exp\left(\frac{\beta}{2} p_1 r\right) R_n^{\alpha+1,\beta}(\tan p_1 r) \tag{17}$$

in equation (16), A and B are constants.

$$\alpha = \frac{n(n-1) - A}{2n-1} \quad ; \quad \beta = \frac{B(2n-1)}{1 + A - n(n+1)} \tag{18}$$

(ii) Proceeding similarly let,

$$\frac{g'^2}{1+g^2} = p_2^2 \tag{19}$$

where p_2^2 is a real positive constant independent of r. Equation (19) gives the functional form of $g(r)$ as $g(r) = \sinh p_2 r$. Using this value in equation (12) yields

$$E_n = -\left[n(n-2\alpha+1) + \alpha(\alpha-1) + \frac{1}{4}\right] p_2^2 \tag{20}$$

$$V(r) = Ap_2^2 \operatorname{sech}^2(p_2 r) + Bp_2^2 \operatorname{sech}(p_2 r) \tanh(p_2 r) - \frac{(D-1)(D-3)}{4r^2} \tag{21}$$

$$\psi(r) = Nr^{-\frac{(D-1)}{2}} (\cosh p_2 r)^{-(\alpha-\frac{1}{2})} \exp\left[\frac{\beta}{2}\tan^{-1}(\sinh p_2 r)\right] R_n^{\alpha,\beta}(\sinh p_2 r) \tag{22}$$

where A, B are constants.

$$A = -\left[\alpha^2 - \frac{1}{4}(1+\beta^2)\right]; \qquad B = -\alpha\beta$$

The third choice $gg'^2/(1+g^2)^2$ in equation (12) is integrable to have $r = r(g)$, but not invertible. The choice g'''/g' equal to constant gives a similar system as in case (ii). The remaining terms $g^2 g'^2/(1+g^2)^2$ and g''^2/g'^2 do not yield potentials of physical interest.

4. Conclusions

We have obtained exactly solvable potentials of radial Schrödinger equation in D-dimensional Euclidean space whose bound state solutions are given in terms of Romanovski polynomials. We have presented a simple transformation method which consists of co-ordinate transformation followed by a functional transformation. The transformation method is performed on the linear second order differential equation satisfied by a particular special function to retrieve radial Schrödinger equation. The constructed potentials are non-power law type with a background inverse square potential $(D-1)(D-3)r^{-2}$ which vanishes for $D = 1$ and $D = 3$. The method can also be applied to construct exactly solvable potentials of Schrödinger equation from other orthogonal polynomials.

References

[1] Bhattacharjie A and Sudarshan E C G 1962 *Nuovo Cimento* **25** 864
[2] Fleases G P 1979 *Phys. Lett.* A **72** 289
[3] Khare A 1981 *Phys. Lett.* A **83** 237
[4] Levai G 1989 *J. Phys. A: Math. Gen.* **22** 689
[5] De R, Dutt R and Sukhatme U 1992 *J. Phys. A: Math. Gen.* **25** L843
[6] Ahmed S A S 1997 *Int. J. Theor. Phys.* **36** 1893
[7] Alvarez-Castillo D E and Kirchbach M 2007 *Rev. Mex. Fis.* E**53** 143
[8] Raposo A P, Weber H J, Alvarez-Castillo D E, Kirchbach M 2007 *Cent. Eur. J. Phys.* **5** 253
[9] Dabrowska J W, Khare A and Sukhatme U P 1988 *J. Phys. A: Math. Gen.* **21** L195
[10] Manning M F 1935 *Phys. Rev.* D **48** 161
[11] Hille E 1969 *Lectures on Ordinary Differential Equations* (Reading, MA: Addison-Wesley) p 647
[12] Khelashvili A A and Nadareishvili T P 2011 *Am. J. Phys.* **79** 668

Warm dust-acoustic shocks

Rajita Goswami[1], **Murchana Khusroo**[1], **and Madhurjya P Bora**[1,2]
[1]Physics Department, Gauhati University, Guwahati 781014, Assam, India.
[2]York Plasma Institute, University of York, York, YO10 5DD, United Kingdom.

E-mail: mpbora@gauhati.ac.in

Abstract. We investigate the formation of shock waves in a warm dusty plasma with equilibrium drift. We show that the natural decay rate of dust charge plays an important role in determining the properties of the shock front in such a plasma.

1. Introduction

We now know that dust grains form an integral part of space and astrophysical plasmas [1–4] and play an important role in determining the dynamical behavior of these systems. This realization led many investigations which reveals the importances of dust grains on various wave modes and their nonlinear counterparts [5, 6]. Rao et. al. [7], are among the first few authors, to study the *dust-acoustic* mode. Numerous investigations have been made which involved the study of evolution of various nonlinear solitary waves [8, 9], shock waves [10, 11], as well as double layers [12, 13]. In this paper, we investigate the dust-acoustic shock waves in a warm dusty plasma.

2. Basic equations of the model

Below are the equations of our 1-D nonlinear model of warm dusty plasma,

$$\frac{\partial n_d}{\partial t} + \frac{\partial}{\partial x}(n_d u_d) = 0, \tag{1}$$

$$m_d n_d \left(\frac{\partial u_d}{\partial t} + u_d \frac{\partial u_d}{\partial x}\right) = e z_d n_d \frac{\partial \phi}{\partial x} - \frac{\partial p_d}{\partial x}, \tag{2}$$

$$\frac{\partial p_d}{\partial t} + u_d \frac{\partial p_d}{\partial x} + \gamma p_d \frac{\partial u_d}{\partial x} = 0, \tag{3}$$

$$\varepsilon_0 \frac{\partial^2 \phi}{\partial x^2} = z_d n_d + n_e - n_i, \tag{4}$$

where $n_{i,e,d}$ are the ion, electron, and dust densities and the other symbols have their usual meanings. While the dust particles are massive and inertial, the electrons and ions are assumed to be Maxwellian, so that the electron and ion densities are solely described by the plasma potential ϕ.

Using a perturbation scheme $f = f_0 + \tilde{f}$, where f_0 is the equilibrium and $\tilde{f}$ is the perturbed parts of any physical quantity f, we can approximate the Poisson equation as,

$$\frac{\partial^2 \tilde{\phi}}{\partial x^2} \simeq \alpha \tilde{\phi} + \beta \tilde{\phi}^2 + \frac{e}{\varepsilon_0} z_d n_d, \tag{5}$$

where

$$\alpha = \frac{e^2}{\varepsilon_0 k_B}\left(\frac{n_{e0}}{T_{e0}} + \frac{n_{i0}}{T_{i0}}\right), \quad \beta = \frac{e^2}{2\varepsilon_0 k_B^2}\left(\frac{n_{e0}}{T_{e0}^2} - \frac{n_{i0}}{T_{i0}^2}\right). \tag{6}$$

The closure of the model will be provided by the dust-charging equation, which is given by,

$$\frac{\partial q_d}{\partial t} = I_d, \tag{7}$$

where q_d is the dust-charge and I_d is the charging current. The dust-charging equation in presence of Maxwellian electron and ions can be simplified as,

$$\frac{\partial \tilde{z}_d}{\partial t} + \eta \tilde{z}_d = p_1 \tilde{\phi} + p_2 \tilde{\phi}^2, \tag{8}$$

where the terms $p_{1,2}$ are given by,

$$p_1 = \frac{|I_{e0}|}{k_B}\left(\frac{1}{T_{io}} + \frac{1}{T_{e0}}\right), \quad p_2 = e\frac{|I_{e0}|}{k_B^2}\left(\frac{1}{T_{eo}^2} - \frac{1}{T_{i0}^2}\right) \tag{9}$$

and η is the natural decay rate of dust-charge [14, 15].

3. Nonlinear perturbation and shock wave solution

We now employ the reductive perturbation to our set of equations, Eqs.(1-3,5,8) by introducing a space and time stretching as $\xi = \epsilon(x - v_0 t), \tau = \epsilon^2 t$, followed by an expansion of the perturbed plasma parameters, $(f = n_d, u_d p_d, \phi), f = \sum_{j=0}^{\infty} \epsilon^j f^{(j)}$, where ϵ is the small expansion parameter. Note that in these expansions, $\phi^{(0)} = 0$. Using this scheme and applying to our model equations, to the lowest order in ϵ, we get the following relations,

$$n_d^{(1)} = -\frac{(\alpha + p/\eta)}{e z_{d0}}\phi^{(1)}, \qquad u_d^{(1)} = -V\frac{(\alpha + p/\eta)}{e n_{d0} z_{d0}}\phi^{(1)}, \tag{10}$$

$$p_d^{(1)} = -\gamma p_{d0}\frac{(\alpha + p/\eta)}{e n_{d0} z_{d0}}\phi^{(1)}, \qquad z_d^{(1)} = \frac{p_1}{\eta}\phi^{(1)}, \tag{11}$$

where $p = e n_{d0} p_1/\varepsilon_0$ and $V = v_0 - u_{d0}$ is the relative velocity of the comoving frame of the perturbation with respecting to the equilibrium drift u_{d0}, that the dust particles might have. The above equations requires a compatibility condition that,

$$V^2 = v_{\text{th}}^2 + \frac{\omega_d^2}{(\alpha + p/\eta)}, \tag{12}$$

where $\omega_d = \sqrt{n_{d0} e^2 z_{d0}^2/(\varepsilon_0 m_d)}$ is the dust plasma frequency and $v_{\text{th}} = \sqrt{\gamma p_{d0}/(m_d n_{d0})}$ is the dust thermal velocity.

Continuing to the next higher order in ϵ, we get the following equations,

$$\frac{\partial n_d^{(1)}}{\partial \tau}+\frac{\partial}{\partial \xi}\left(n_d^{(1)}u_d^{(1)}\right) = V\frac{\partial n_d^{(2)}}{\partial \xi}-n_{d0}\frac{\partial u_d^{(2)}}{\partial \xi}, \tag{13}$$

$$n_{d0}\left(\frac{\partial u_d^{(1)}}{\partial \tau}+u_d^{(1)}\frac{\partial u_d^{(1)}}{\partial \xi}\right) = V\left(n_d^{(1)}\frac{\partial u_d^{(1)}}{\partial \xi}+n_{d0}\frac{\partial u_d^{(2)}}{\partial \xi}\right)+\frac{ez_{d0}}{m_d}\left(n_{d0}\frac{\partial \phi^{(2)}}{\partial \xi}+n_d^{(1)}\frac{\partial \phi^{(1)}}{\partial \xi}\right) + \frac{en_{d0}}{m_d}z_d^{(1)}\frac{\partial \phi^{(1)}}{\partial \xi}-\frac{1}{m_d}\frac{\partial p_d^{(2)}}{\partial x}, \tag{14}$$

$$-\frac{\partial p_d^{(1)}}{\partial \tau} = u_d^{(1)}\frac{\partial p_d^{(1)}}{\partial \xi}+\gamma p_d^{(1)}\frac{\partial u_d^{(1)}}{\partial \xi}+p_{d0}\frac{\partial p_d^{(2)}}{\partial \xi}-V\frac{\partial p_d^{(2)}}{\partial \xi}, \tag{15}$$

$$0 = zn_d^{(2)}+n_d^{(1)}z_d^{(1)}+n_{d0}z_d^{(2)}-\beta\left(\phi^{(1)}\right)^2-\alpha\phi^{(2)}, \tag{16}$$

$$0 = \eta z_d^{(2)}-p_2\left(\phi^{(1)}\right)^2-p_1\phi^{(2)}-v_0\frac{\partial z_d^{(1)}}{\partial \xi}. \tag{17}$$

Eliminating the second order quantities from Eqs.(13-17) and using the results for the first order quantities in terms of $\phi^{(1)}$, we finally derive the nonlinear shock wave equation, known as the Burger equation [16] as,

$$\frac{\partial \phi^{(1)}}{\partial \tau}+A\phi^{(1)}\frac{\partial \phi^{(1)}}{\partial \xi}+B\frac{\partial^2 \phi^{(1)}}{\partial \xi^2}=0, \tag{18}$$

where the coefficients are given by,

$$A = \frac{(\alpha+p/\eta)}{2Vn_{d0}}\left[2V^2+v_{\rm th}^2(\gamma-1)\right]+\frac{(V^2-v_{\rm th}^2)}{Vz_{d0}}\left[\frac{1}{(\alpha+p/\eta)}\left(\frac{p_2n_{d0}}{\eta}+\varepsilon_0\alpha\right)+\frac{p_1}{\eta}\right] -\frac{e}{2m_dV}\left[\frac{p_1n_{d0}}{\eta(\alpha+p/\eta)}+z_{d0}\right], \tag{19}$$

$$B = \frac{v_0n_{d0}(V^2-v_{\rm th}^2)}{2(\alpha+p/\eta)V\eta^2z_{d0}}. \tag{20}$$

The coefficient A is the coefficient of nonlinearity and B represents dissipation. Although, our model does not include any physical dissipation, the dust-charge fluctuation is manifested as dissipation in our model, which drives the shock front.

4. Concluding remarks

We note that the natural decay rate of dust charge η plays an important role in determining the behaviour of the shock front. Taking an asymptotic limit of η, on the dissipation coefficient B, we get,

$$B \simeq \frac{v_0\varepsilon z_{d0}}{2m_dp_1\sqrt{\omega_d^2\alpha^{-1}+v_{\rm th}^2}}, \quad \text{for } \eta\simeq 0, \qquad B\simeq 0, \quad \text{for } \eta\gg 1, \tag{21}$$

which reaches a constant value for small η and become zero for large η. This essentially means that when the natural decay rate of dust charge becomes zero, the shock front still survives owing to the dust-charge fluctuation whereas when η becomes large, the dust-charge dissipates very quickly, which destroys the shock front. It is interesting to look at these asymptotic values when the dust is essentially cold i.e. for small p_{d0},

$$B \simeq \frac{\varepsilon_0v_0}{2ez_{d0}}\sqrt{\frac{\omega_d^2}{p\eta}}, \quad \text{for } \eta, p_{d0}\simeq 0, \tag{22}$$

Journal of Physics: Conference Series **481** (2014) 012028 doi:10.1088/1742-6596/481/1/012028

when B becomes very large for small η.

References

[1] C. K. Goertz, Phys. Scr. 27, (1989) 271.
[2] T. G. Northrop and T. J. Birmingham, planet. Space Sci. 38, (1990) 319.
[3] F. Melandso, T. K. Aslakson, and O. Havnes, Planet. Space Sci. 41, (1993) 321.
[4] N. D' Angelo, Planet. Space Sci. 42, (1994) 507.
[5] C. B. Dwivedi and B. P. Pandey, Phys. Plasmas 2, (1995) 4134.
[6] N. N. Rao and P. K. Shukla, Planet. Space Sci. 42, (1995) 221.
[7] N. N. Rao, P. K. Shukla, and M. Y. Yu, Planet. Space Sci. 38, (1993) 543.
[8] F. Sayed and A. A. Mamun, Phys. Plasmas 14, (2007) 014502.
[9] M. A. Mirza, S. Mahmood, N. Jehan and N. Ali, Phys. Sci. 75, (2007) 755.
[10] B. Eliasson and P. K. Shukla, Phys. Rev. E 69, (2004) 067401.
[11] P. K. Shukla and A. A. Mamun, New J. Phys. 5, (2003) 173.
[12] M. A. Raadu, Phys. Rep. 78, (1997) 25.
[13] G. C. Das and K. Devi, Astrophys. And Space Sci. 330, (2010) 79.
[14] G. C. Das, Balen Choudhury, and M. P. Bora, Phys. Plasmas 17, (2010) 123707.
[15] M. R. Jana, A. Sen, and P. K. Kaw, Phys. Rev. E 48(5), (1993) 3930.
[16] J. M. Burger, Proc. R. Neth. Acad. Soc. 32, (1929) 414.

AUTHOR INDEX

Adhikari, Rathin 75
Barua, A. G. 59
Batra, M. 103
Bhagawati, Nabaratna 115
Bhattacharjee, Anindita 55
Bora, Kalpana 22, 71, 83
Bora, Madhurjya P. 119
Borah, Neelakshi N. K. 99
Chakraborty, Sarani 39
Chakraborty, Subrata 13
Choudhury, D. K. 1, 87, 91, 95, 99, 107
Dasgupta, Arnab 75
Datta, Anindya 28
Deka, Kabita 43
Deshamukhya, Atri 55
Devee, Mayuri 111
Dey, Ujjal Kumar 28
Dhar, Pijush Kanti 91
Dubey, Anuj Kumar 47
Duorah, H. L. 59
Duorah, K. 59
Dutta, Debajyoti 83
Garg, Sumit K. 79
Ghosh, Gayatri 71
Goswami, Deepanjali 79
Goswami, Rajita 119
Goswami, Umananda Dev 43
Jahan, Akbari 107
Khusroo, Murchana 119
Medhi, Jayashri 59
Pathak, Krishna Kingkar 95
Pathak, S. D. 67
Poulose, P. 79
Rahman, Zini 75
Raychaudhuri, Amitava 28
Roy, Sabyasachi 87
Sahariah, P. K. 99
Sarma, J. K. 111
Sen, Anjan A. 18
Sen, Asoke Kumar 39, 47
Shaw, Avirup 28
Singal, Ashok K. 34
Singh, Dharm Veer 63
Singh, N. Nimai 6
Siwach, Sanjay 63
Upadhyay, A. 103
Verma, M. M. 51, 67

Institute of Physics
Dirac House, Temple Back
Bristol BS1 6BE UK

CURRAN ASSOCIATES INC
proceedings
.com